LANDMARK COLLECTOR'S LIBRARY

Stationary Steam Engine Makers

Volume 1

Compiled by George Watkins

Catalogued by A. P. Woolrich

The Watkins Collection in the National Monuments Record

This comprises the photographs and notes George Watkins made during a lifetime of study of the stationary steam engine.

The Steam Engine Record is an annotated set of around 1,500 mounted prints of steam engines (and thousands of others) which Watkins examined in the field between 1930 and 1980. His notebooks contain a record of additional sites for which no photographs were taken, or which comprise written historical notes.
In all almost 2000 entries were made in his notebooks.
There are also albums of prints arranged by engine type. A catalogue is available.

In addition there are files of notes and other records on all aspects of historical steam technology, the cataloguing of which is in progress.

The main areas of this part of the collection are:

**Records of steam engine makers.*

**Collection of bound trade literature.*

**Classified collection of data files dealing with, for example, textile mill engines, marine engines.*

The original material may be consulted at the National Monuments Record, the public archive of English Heritage. For further information please contact.

NMR Enquiry & Research Services
National Monuments Record Centre
Kemble Drive, Swindon SN2 2GZ

email: nmrinfo@english-heritage.org.uk
www.english-heritage.org.uk/nmr

LANDMARK COLLECTOR'S LIBRARY

STATIONARY STEAM ENGINE MAKERS

VOLUME 1

Compiled by George Watkins

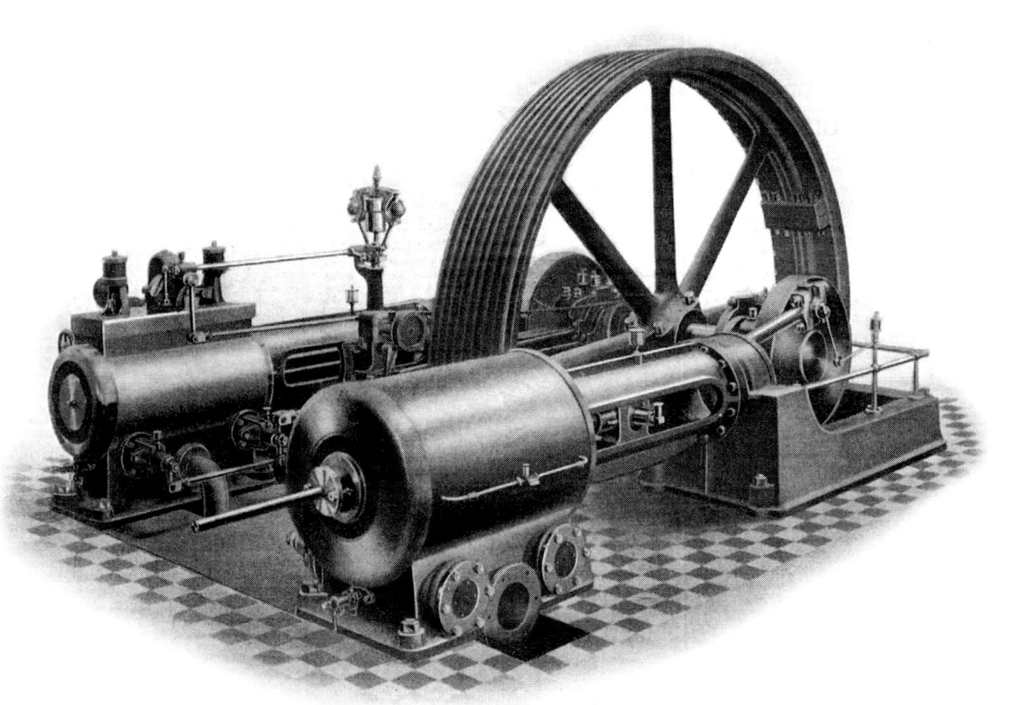

Catalogued by A. P. Woolrich

Landmark Publishing

Published by

Ashbourne Hall, Cokayne Ave
Ashbourne, Derbyshire, DE6 1EJ England
Tel: (01335) 347349 Fax: (01335) 347303
e-mail: landmark@clara.net
web site: www.landmarkpublishing.co.uk

ISBN 1-84306-200-3

British Library Cataloguing in Publication Data: a catalogue record for this book is available from the British Library.

Print: Bath Press Ltd

Design & reproduction by: James Allsopp

Production: C. L. M. Porter

Front cover: Hick, Hargreaves

Back cover top: Fullerton

Back cover bottom: Browett

Page 3: Easton & Bessemer

Contents

INTRODUCTION

By A. P. Woolrich

This two volume book is based on the catalogue of 375 stationary steam engine makers' files comprising section 34 of the Watkins Collection at English Heritage's National Monuments Record, Swindon. The materials covered are the files Watkins prepared himself and the cataloguing was done in the early 1990s when the collection was housed at the Royal Commission for Historic Monuments' office at Salisbury. The collection is now in English Heritage's National Monuments Record at Swindon.

These books should be used in conjunction with Section 35 of the Watkins Collection, a 21- volume bound series of trade catalogues classified by topic.

A summary contents listing there, as well as the list sorted by maker is included in the companion volume by A. P. Woolrich, *Stationary Steam Engine Makers – a reference catalogue from the Watkins Collection*. This includes other reference indices from trade journals, and a series of short company histories, both compiled by Dr Martin Doughty, and material about a survey Watkins made in 1966 of surviving steam engine makers in Britain and America.

Watkins was an assiduous collector of information about stationary steam engine manufacturers world-wide, and his files contain a great deal of information about famous makers as well as obscure ones. The files in the archive at Swindon are in alphabetical order, with a numerical code for each one.

For a large number of concerns mentioned there is very little recorded as they were defunct before Watkins began his work. Sometimes an engine of theirs might only be in the Steam Engine Record of photographs, or there may be a couple of advertisements in the files. Perhaps there is a note about their activities from an engineering periodical. In a minority of concerns only is there anything more comprehensive, and these relate to major engine builders. Even here the coverage is not a full one, since there are several very well-known concerns like Maudslays, or James Watt & Co., who are either very sparsely covered or are missing completely from the files of advertisements and other papers. Just why this is so has not yet been discovered. It is possible that the files passed into other hands for some reason and were not returned. Alternatively the material might be retained with the major subject areas like mining, marine engines or textiles, though this is not obvious from the file headings actually used for such sections.

Watkins's interest in engine makers was international in scope, but most of the foreign material came from North America and Canada. There is some material from German, Dutch and Belgian concerns, with a few items from French and Czech ones. There is a little from Australia and New Zealand. He began collecting catalogues from firms that were still making engines in the 1920s and these form the bulk of the material in the bound series.

A great many of the makers listed in these volumes are illustrated with a selection of images from the advertisements, photographs and trade literature. While they are listed here, what are not included are many images showing engines being manufactured or erected. These are being used to illustrate a companion volume about stationary engine manufacture and construction.

Each file was sorted into the following classification, but not every maker's entry has data for every classification, so in order to keep these books concise, classifications where there are no data have been ignored.

Advertisements

The advertisements are perhaps the most important part of the material, since they go back to the later decades of the nineteenth century, but it is not usually possible to identify which periodical or issue they came from. Watkins annotated them with the year of publication, however. He clipped them from the wrappers which accompanied the various technical journals as they were published weekly. When he was active, particularly before World War II, it was possible to buy back-numbers of individual issues from the publishers' offices in London for 6d each, and he was able to clip the advertising pages for these files and break up the editorial texts for filing elsewhere in his collection. He also preserved a selection of complete issues with wrappers for each decade from the 1860s to the turn of the century. Journals like *The Engineer* are to be found today in libraries in bound volumes, with the wrappers discarded, so this collection is a unique resource for technical historians.

For a period the advertisement pages of some journals were gridded, so that the advertising copy appeared exactly at the same positions on the fronts and backs of the pages. In the pages which follow, where a maker's advertisements are listed they are in date order, and are followed by an entry recording the details of the advertisement on the back. These 'reverse' entries, with 327 names, are indexed at the end of the book, providing a finding aid for information about engineering companies which did not always make steam engines.

A depressing feature is the instances where the only material about a company is one clipping recording the bankruptcy or sale of the works.

Letters

Many of the files include letters to and from Watkins as he undertook his research. He was assiduous in enquiring about the fate of makers' records, and it is regrettable to read how many collections were destroyed just before he enquired. He was an avid correspondent with fellow historians and their letters often contain useful information.

MS material

These are usually notes Watkins wrote about the histories of the firms.

Photographs

A number of the files have valuable collections of photographs. These are not the same as the photographs Watkins took himself and which constitute the Steam Engine Record. A number are official works' photographs, and others are photographs correspondents sent to him.

Printed material

These comprise articles torn from the technical press, or photocopies of them.

Trade literature

These comprise trade catalogues and leaflets published by the firms concerned.

Other material

Literature not covered by the previous heads are found here.

Readers are also referred to the alphabetical list of makers in the final volume of Landmark Publishing's series George Watkins, *The Stationary Steam Engines of Great Britain.* 10 vol.

In the search room at Swindon is a Consolidated Makers' Index based on this one which includes extra information about each maker gathered from different parts of the collection. The sheets from this bigger index for each maker appear in the maker's envelope (section 34 of the Collection).

A. P. Woolrich

Selection of illustrations

Care has been taken to choose images which are clear, with a preference for older ones, where the date is known. More recent images have not been chosen to avoid copyright issues. It is believed that the illustrations are free of copyright, or where it possibly existed, an attempt to obtain permission has been made.

Clearly, the selection was limited by what exists and for a lot of smaller manufacturers, the record may only be an advertisement. The diversity of the documentation is its chief strength, including such items as complete lists of engines made and other work by researchers who sent a copy to George Watkins. Another is the quantity of technical press features on engines. While these are fascinating in their own right, the text has not been chosen for inclusion, although some drawings accompanying such features have been included.

A lot of subjectivity has gone into the inclusion, but when completed, it left a feeling that it is probable that no other publication includes such a diversity of engine makers' products. It certainly gives a good idea on what the Collection includes and complements the astonishing detail of the catalogue entries as a whole.

Catalogue Entries

These are prefixed by a maker's number, starting with MAK/001. If you wish to order copies of any files, simply quote the name and maker's number.

Lindsey Porter
Landmark Publishing

MAK/001 - Daniel Adamson, Newton Moor Ironworks, Dukinfield, nr Manchester.

1) Advertisements:
1880, 1885, 1894 (2), No date.

Reverse: 1894, Rose, Downs and Thompson, Old Foundry, Hull.

Reverse: No date. Laurence, Scott & Co., Gothic Works, Norwich on electric motors.

2) Letters:
From Miss A. Teague, Dukinfield, 16 April 1958 about history of the firm and old catalogues.

From F. Buckley, Dukinfield 10 June 1958 on his time at Adamsons, c 1908.

5) Printed material:
Account of visit to works by the Institution of Mechanical Engineers, 1875.

Odd pages and plates from Adamson's paper on HP steam and QE engines from *Jnl, Iron and Steel Institute*, 1875, photocopies.

Article ex *Engineering* 4 Jan 1876 (?1867, see p.14), pp 4, 9 on a horizontal compound engine [plate only] and Adamson's boiler works, Newton Moor, nr Manchester.

Photocopies of an article ex *Engineering,* 14 April 1876, pp 290-292. on Adamson's works at Hyde Junction.

Article ex *Mechanical World*, 16 October 1884, pp 264-266 on Wheelock expansion gear for steam engines, made by Adamsons.

Article ex *Engineering*, 24 March, 1893, pp 350-359 on 1500 HP triple expansion mill engine for Minerva spinning mills, Ashton under Lyme [*Capital* and *Labour*] made by Adamsons.

Article ex *Mechanical World,* 21 April, 1893 on previous engine.

Article ex *The Engineer* 10 April, 1896, p 375 on engine for Dowlais Iron Works, made by Adamsons.

6) Trade literature:
Catalogue of machinery No 3, 1869 16pp.

Catalogue of steam engines, 32pp n.d., but c 1910.

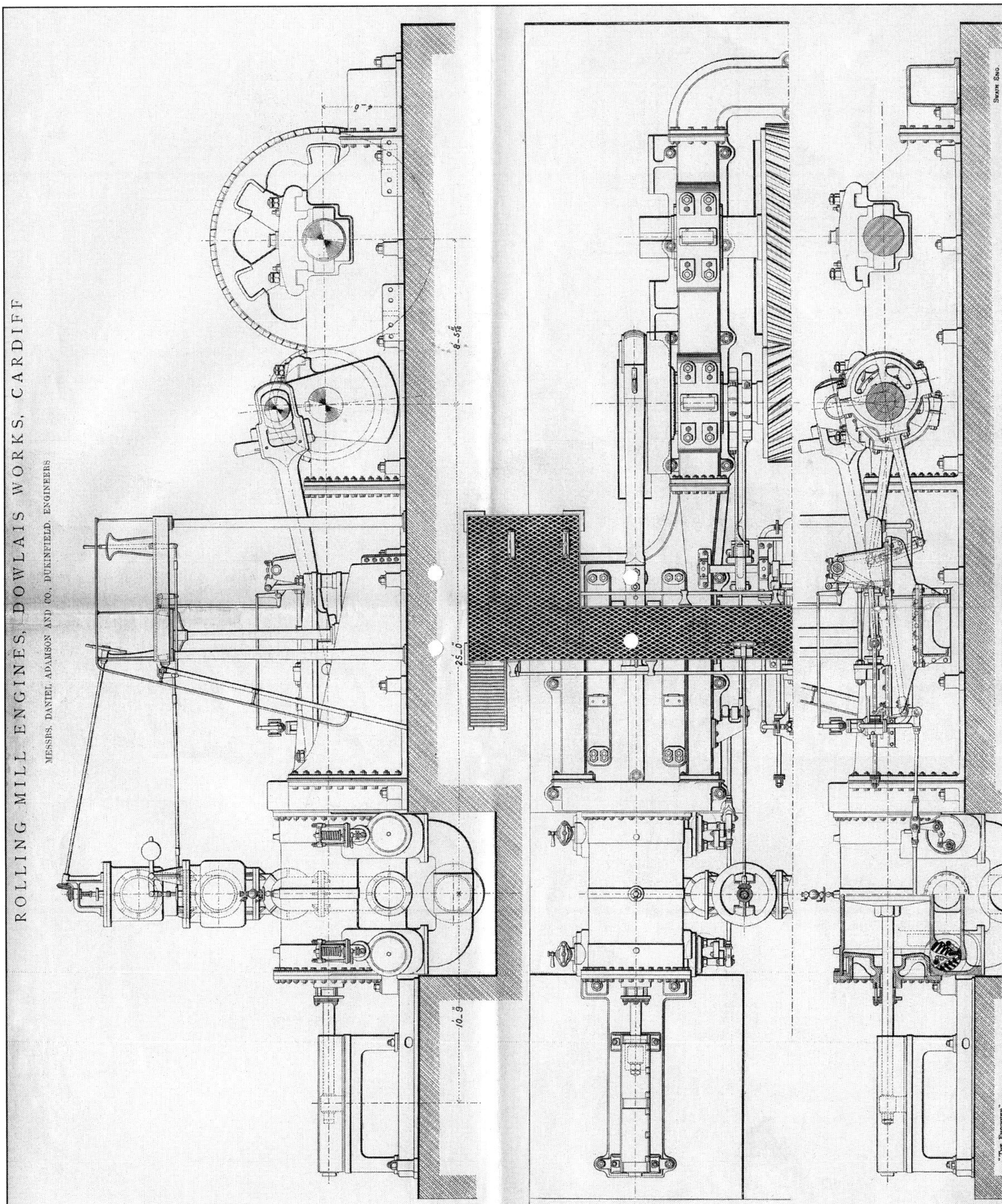

The *Engineer* 10/4/1896, p375

HORIZONTAL COMPOUND STEAM ENGINES.

CONSTRUCTED BY DANIEL ADAMSON AND CO., ENGINEERS, HYDE.

(For Description, see preceding Page.)

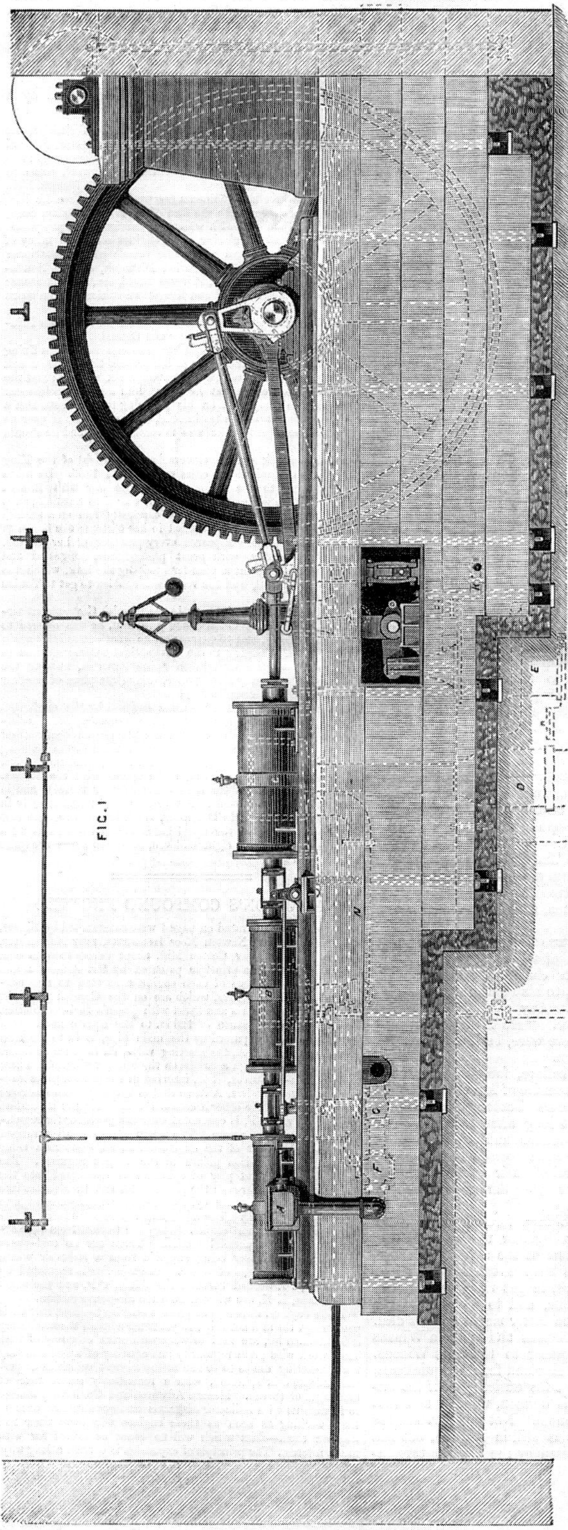

FIG. 1

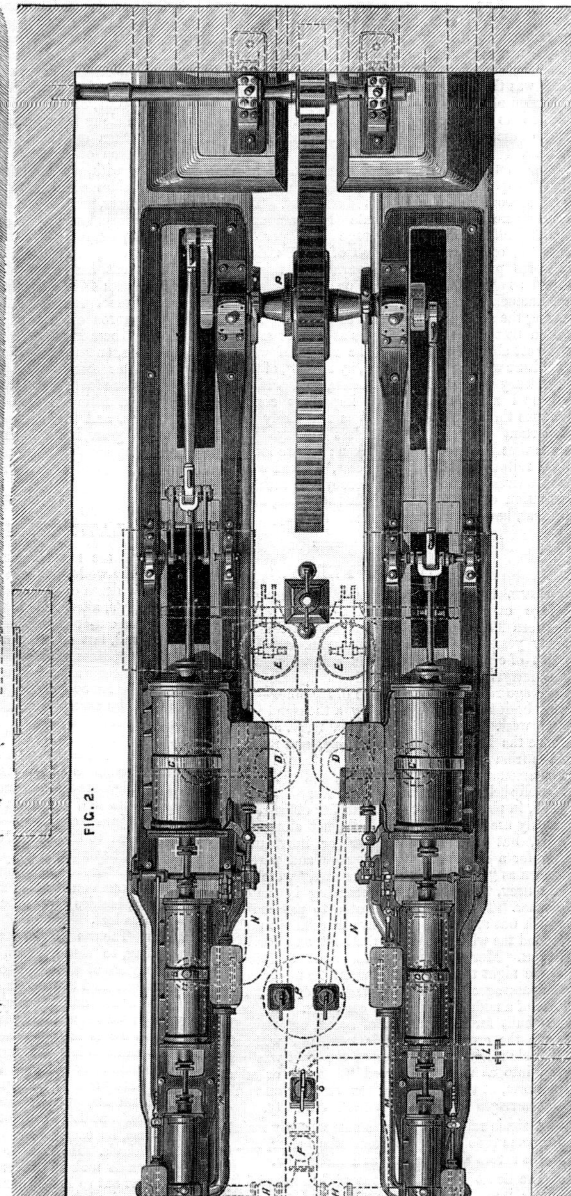

FIG. 2

MAK/002 - Joseph Adamson, Hyde, nr Manchester.

1) Advertisements:
1885, no date.

2) Letters:
From Joseph Adamson & Co., 4 August 1959 about boilers at Dee Mill, Shaw, Lancs.

MAK/003 - Aimers, McLean & Co., Waverley Iron Works, Galashiels Scotland.

1) Advertisements:
Large leaflet advertising an improved wall engine, 1-16hp, single leaf, no date nor address, 4 identical copies (see below).

2) Letters:
From Aimers, McLean 28 July 1961 about making GW a copy of a drawing.

AIMERS, M'LEAN & COY.

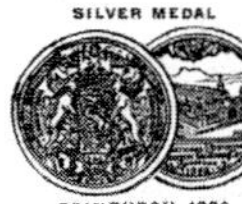

HIGH PRESSURE AND CONDENSING.

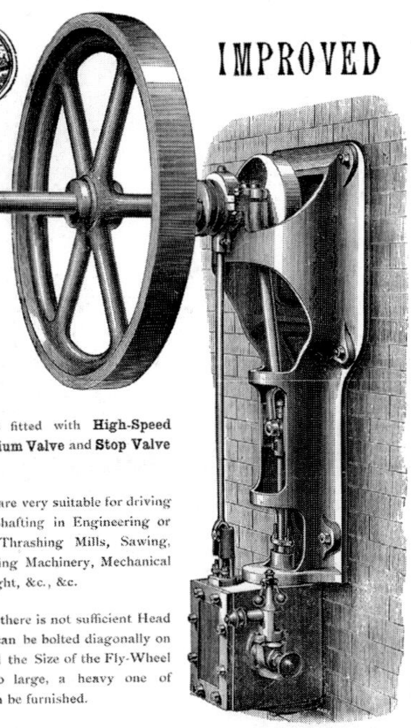

IMPROVED WALL ENGINE.

THE
Cheapest in First Cost
AND
SECOND TO NONE FOR SIMPLICITY, STRENGTH, DURABILITY,
AND
Superior Workmanship.

REQUIRES NO FOUNDATION,
as it can be bolted to any ordinary Wall; and, being self-contained, causes **no vibration** or injury to building.

Saves Belts, Pulleys, or Gearing—being coupled direct to main Shaft.

Takes Less Time to Erect.

The Engine is fitted with **High-Speed Governor, Equilibrium Valve** and **Stop Valve** combined.

These Engines are very suitable for driving separate lines of Shafting in Engineering or other Works, for Thrashing Mills, Sawing, Pumping and Hoisting Machinery, Mechanical Stokers, Electric Light, &c., &c.

NOTE.—Where there is not sufficient Head Room the Engine can be bolted diagonally on the wall; or should the Size of the Fly-Wheel given below be too large, a heavy one of smaller diameter can be furnished.

The **Bed-Plate, Top Cylinder Cover, Cross-head Guides, and Plumber Block** for **Crank Shaft Bearing** are cast in one piece, the **Cylinder** being bolted to the bottom of the **Bed-Plate** and neatly cased with sheet iron. The **Cross-head Slippers** and **Connecting Rod Ends** are made adjustable so that any wear can readily be taken up.

The **Fly-Wheel** is turned bright and rounded on the face so that it can be used for **Driving Belt**.

NOMINAL HORSE POWER.	4	6	8	10	14	16
	Ft. In.	Ft. In.	Ft. In.	Ft. In.	Ft. In.	Ft. In.
Diameter of Cylinder, ...	6	8	9	10	12	14
Length of Stroke, ...	12	16	18	18	20	20
Diameter of Steam Inlet, ...	1¼	1½	1¾	2	2½	3
Diameter of Exhaust Outlet, ...	1¾	2¼	2½	3	3½	4
Diameter of Fly-Wheel, ...	4 0	4 6	5 0	5 0	6 0	6 0
Width of Face (Turned), ...	5	6	7	8	9	10
Diameter of Crank Shaft, ...	2¾	3¼	3½	3¾	4¼	4½
Length of Crank Shaft, ...	4 0	4 6	5 0	5 6	6 0	6 6
Distance from Bottom of Cylinder to Top of Bed-Plate,	6 3	8 2	9 2	9 2	10 3	10 3
Extreme Width of Bed-Plate, ...	2 2	3 0	3 4	3 4	3 9	3 9
Distance from bottom of Cylinder to Centre of Crank Shaft, ...	5 7	7 2	7 11	7 11	8 11	8 11
Revolutions per Minute, ...	150	140	130	120	110	100
Approximate Weight of Engine, ...	12 cwt.	25 cwt.	37 cwt.	38 cwt.	60 cwt.	72 cwt.
Price of Engine with Improved High Speed Governor, but no Feed Pump, ...						
Bolts and Plates for Bolting Engine to Wall, *Extra*,						

Packing Cases Extra.

MAK/004 - Airdrie Iron Co., Airdrie Iron Works, Airdrie, nr Glasgow.

1) Advertisements:
1891.

5) Printed material:
Extract from *Mechanical World,* 26 May 1888, describing engines exhibited at Glasgow Exhibition.

MAK/005 - Aitchison Blair, Clydebank.

2) Letters:
Letter to GW from Bill Lind, 'Throsk', Kilbarchan Road, Bridge of Weir, Renfrewshire, PA1 1 3EZ, 9 January 1973. Concerns closure of the firm and the recovery by Lind of the entire business records, as well as two complete sets of their drawings. Lind says he was still trying to locate two boxes of glass plate negatives representing engines made over 60 years.

MAK/006 - Alexander & Son, Cirencester, Glos.

1) Advertisements:
1875.

MAK/007 - William Allchin, Globe Works, Northampton.

1) Advertisements:
1882 (2).

MAK/008 - Alldays & Onions, Birmingham.

1) Advertisements:
1893.

MAK/009 - W. H. Allen of Lambeth, London, and Queen's Engineering Works, Bedford.

1) Advertisements:
188(?), 1891 (see p.17), 1893, 189(?), 1900, 1903, 1905.

Reverse: 188(?) The Falcon Engine and Car Works, Ltd. 41 Coleman Street, London and Loughborough.

Reverse: 1891, Priestmann oil engine (cropped).

Reverse: 1900 Blake-Knowles Steam Pump works (address not given).

6) Trade literature:
Booklet: about industrial service plant, 20pp, n. d. but c 1950.

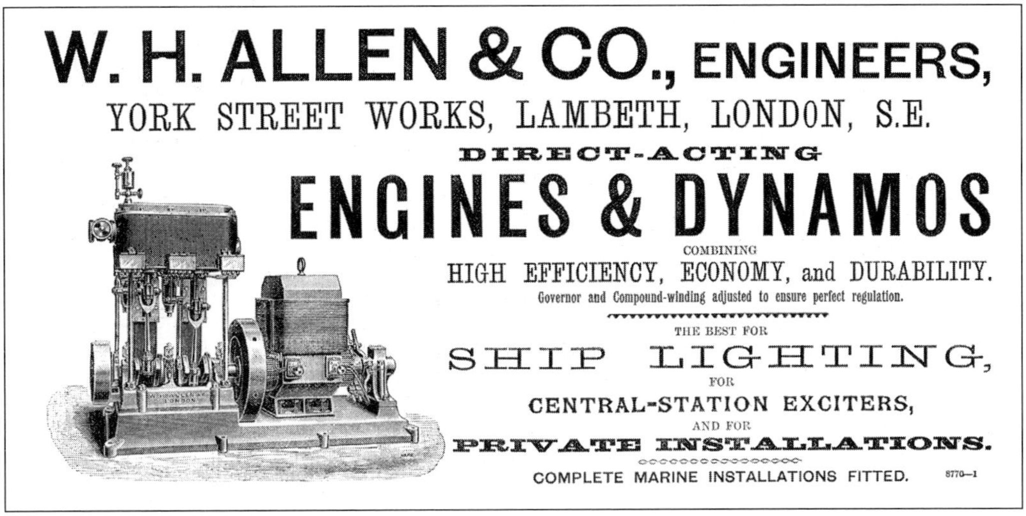

MAK/010 - Alley & MacLellan, Sentinel Works, Glasgow.

1) Advertisements:

1887 (3), 1894, 1900, 1901

Reverse: 1887 Yarrow, London stern wheelers.

Reverse: 1894 Clarke, Chapman & Co. Gateshead on Tyne. Electrical engineers, Portable electric dip pump.

Reverse: 1900 Robert Warner & Co., 97 Queen Victoria Street, London - boring machinery; John Fraser, Millwall Boiler Works, London - evaporative condensers.

Reverse: 1901 Thomas Shanks, Union Iron Works, Glasgow - heavy machine tools.

6) Trade literature:

Series of pages torn from catalogue, n.d. but c 1920, pp 5-6, 21-28 43-48, and pp 145 -148 of Catalogue: 22-S.

Photocopy of pp 30, 31 and 36 of perhaps previous catalogue:, with pp 44, 45, 48, 49 of Catalogue: no 45, Air.

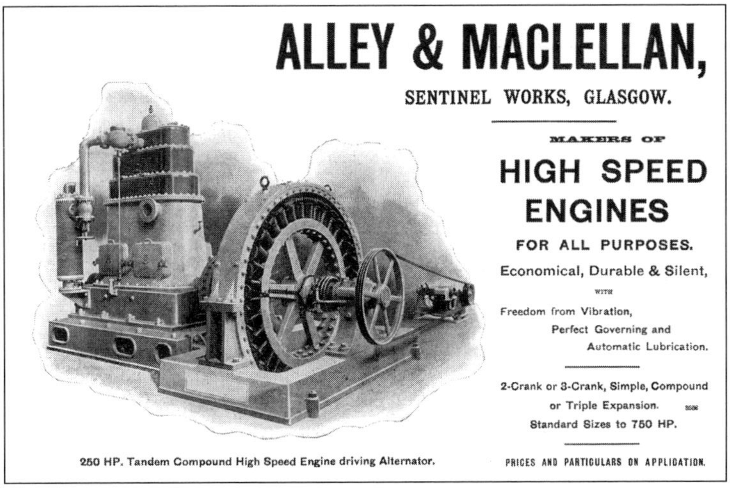

250 HP. Tandem Compound High Speed Engine driving Alternator.

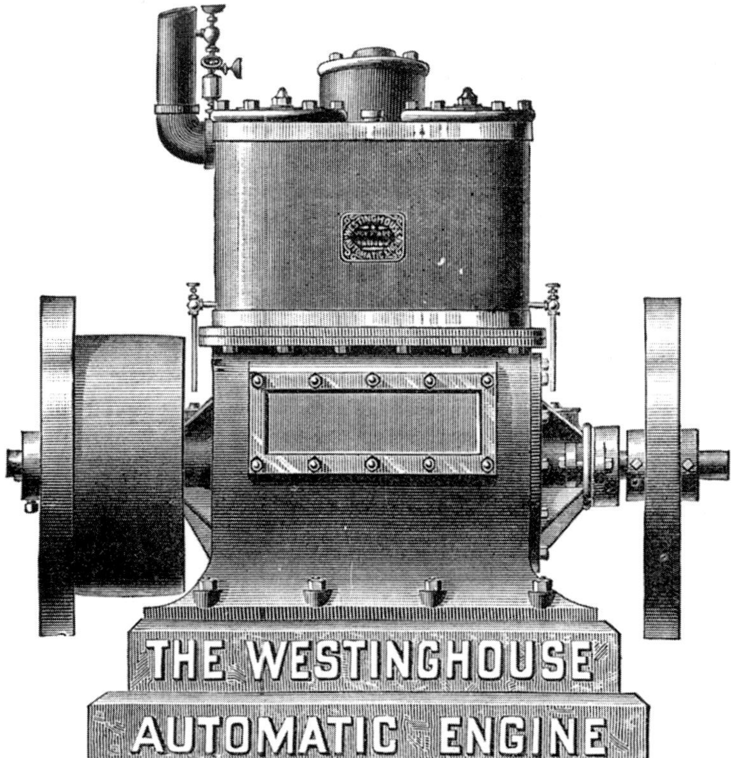

SOLE MAKERS FOR EUROPE, INDIA AND THE COLONIES,

ALLEY & MACLELLAN,

SENTINEL WORKS, GLASGOW.

49, QUEEN VICTORIA STREET, LONDON, E.C.

—WRITE FOR CIRCULAR.—

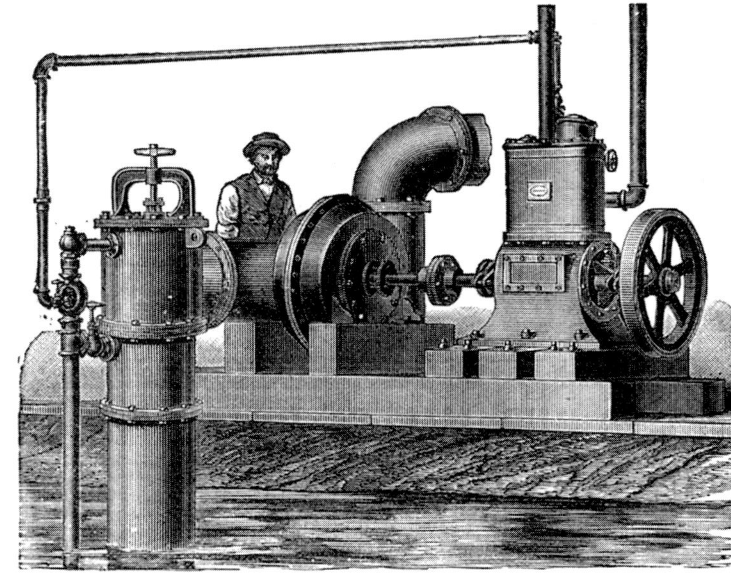

WESTINGHOUSE ENGINE, CONNECTED DIRECT TO A CENTRIFUGAL PUMP, FOR DRAINAGE IRRIGATION, &C.

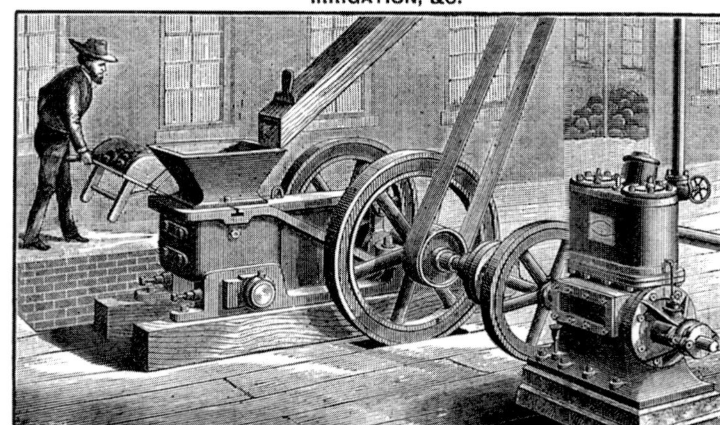

6481

WESTINGHOUSE ENGINE, CONNECTED DIRECT TO AN ORE OR STONE CRUSHER.

MAK/011 - Allis-Chalmers, Milwaukee, Wis., USA.

1) Advertisements:
1900 [E. P. Allis], 1903 (2), 1905
Reverse: 1900 Lytham Shipbuilding and Engineering Co., Lytham.
Reverse: 1903 D. Stewart & Co., (1902) Ltd, London Road Iron Works, Glasgow- slow speed engines; L. Sterne & Co., Crown Ironworks, Glasgow - refrigeration and icemaking machinery.
Reverse: 1905 Urquhart, Lindsay & Co., Blackhouse Foundry, Dundee, for cast iron wheels.

2) Letters:
From Allis-Chalmers, 26 Aug 1966, with list of their engines sold to British concerns, 1897-1902, with useful incidental detail.

From Allis-Chalmers, 2 May 1967, with copies of some of the pages from catalogue in 6 below.

4) Photographs:
Collection of 8 unidentified prints of various engines. All look like reproductions from catalogues. [Some numbered, the highest no 14, so there may be more elsewhere in the collection.].

6) Trade literature:
Illustrated catalogue: of engines, n.d. but c 1900, 93pp, landscape format. Many fine woodcuts. Cover disintegrated with acid rot. (see p.20).

Booklet: about works and products, 3ed, 1925, 66pp.

Photocopy of list of Corliss engines, totalling over five million hp, 1914, 82pp.

Photocopy of illustrated spare-parts list for Corliss engines, 1919 8pp.

Photocopies of odd pages of Sales Bulletin, Aug, 1920, Aug 1922, Jan 1925, Jan 1932.

Photocopy of title page to list of over 300,000 hp of gas engines [NB Perhaps rest is in GW's Gas Engines files. APW].

7) Printed material:
Photocopy of TS *Allis-Chalmers Corporation - A brief history,* 4pp. n.d., but post 1974.

Photocopy of works drawing lists for Corliss engine No 230, 24in x 40in x 42in, and No 301 18in x 28in x 36in, both made for Honolulu Ironworks, 1927-28.

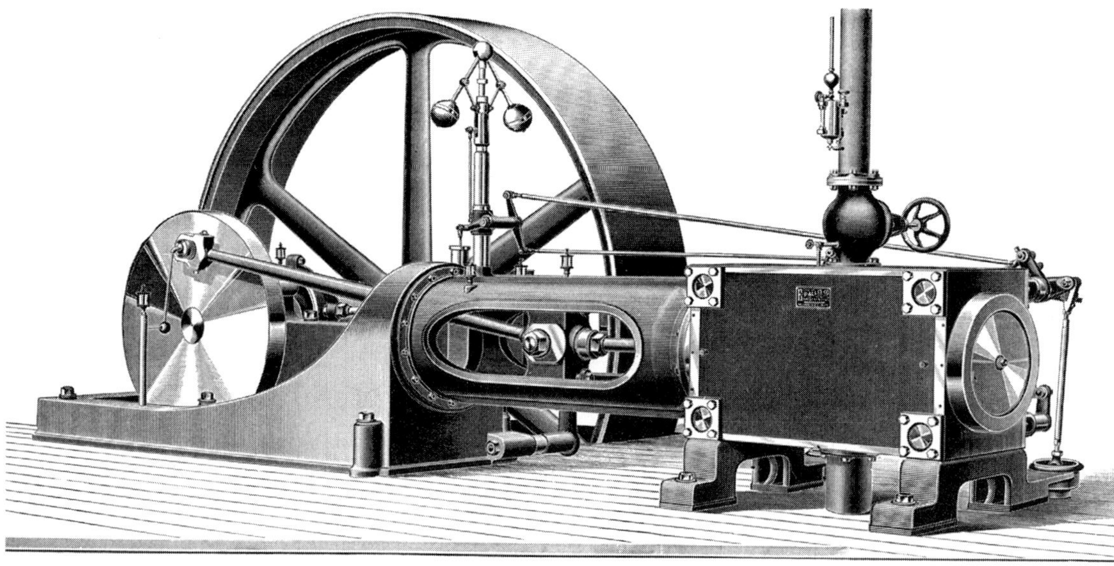

REYNOLDS' CORLISS "1890" ENGINE.—FRONT VIEW.

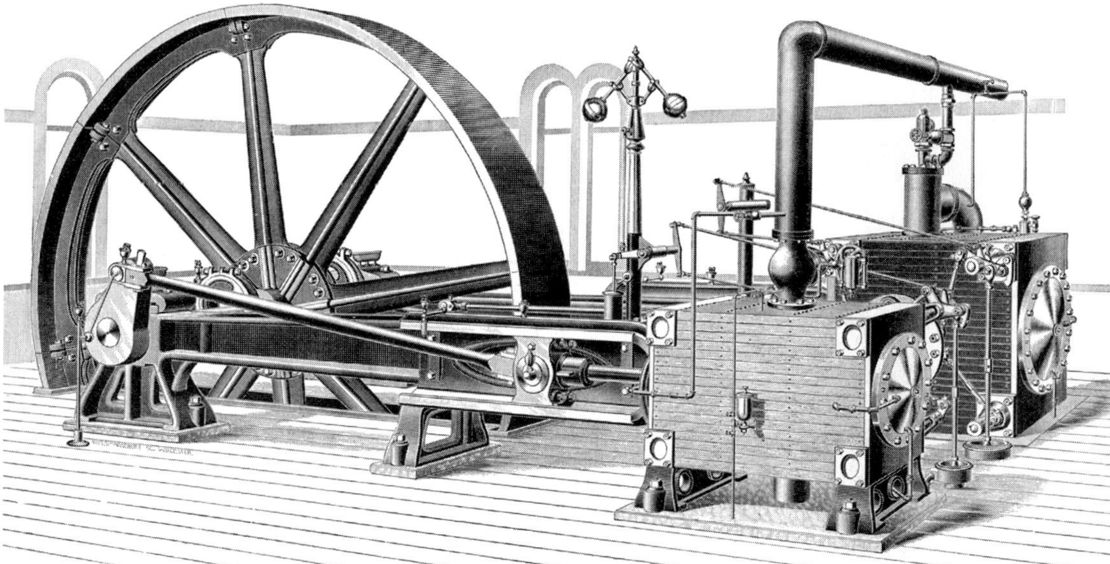

CROSS COMPOUND CONDENSING ENGINE.

MAK/012 - William Allsop & Co., Preston.

1) Advertisements:
1896 (sale of works).

MAK/013 - Ames, Oswego, New York, USA.

6) Trade literature:
Catalogue:, *The Unaflow Engine*, 1924, 32pp.

MAK/014 - Anderston Foundry Co., Glasgow & Middlesbrough.

1) Advertisements:
Steam generating plant.

MAK/015 - Anderton & Sons, Antley Boiler Works, Accrington.

1) Advertisements:
1895, 1897, 1906 (see below), 1908.
Reverse: 1895 Lowood - Ganister bricks; Metallic Valve Co., Tower Buildings, Water Street, Liverpool.

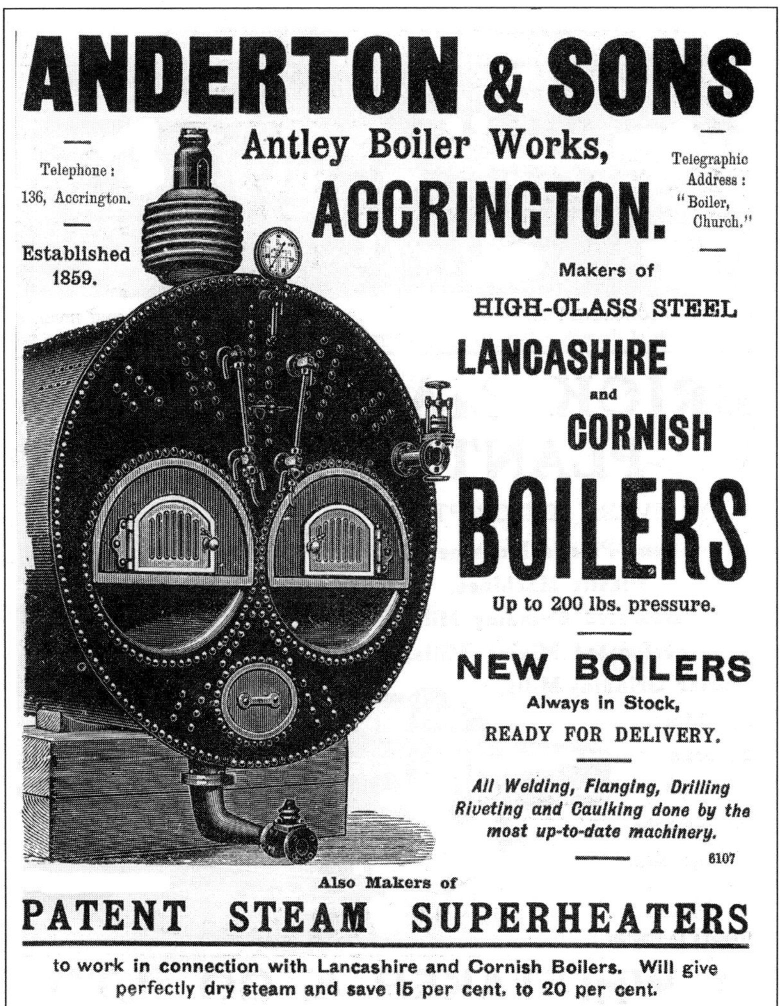

MAK/016 - Appleby Crane & Transporter Co., Grafton Place, St Margaret's, Leicester.

1) Advertisements:
2 copies of advertisement dated 16 June 1911 about the sale of the works.

MAK/017 - Sir William Arroll & Co., Ltd, Dalmarnock Ironworks, Bridgeton, Glasgow.

1) Advertisements:
1905.

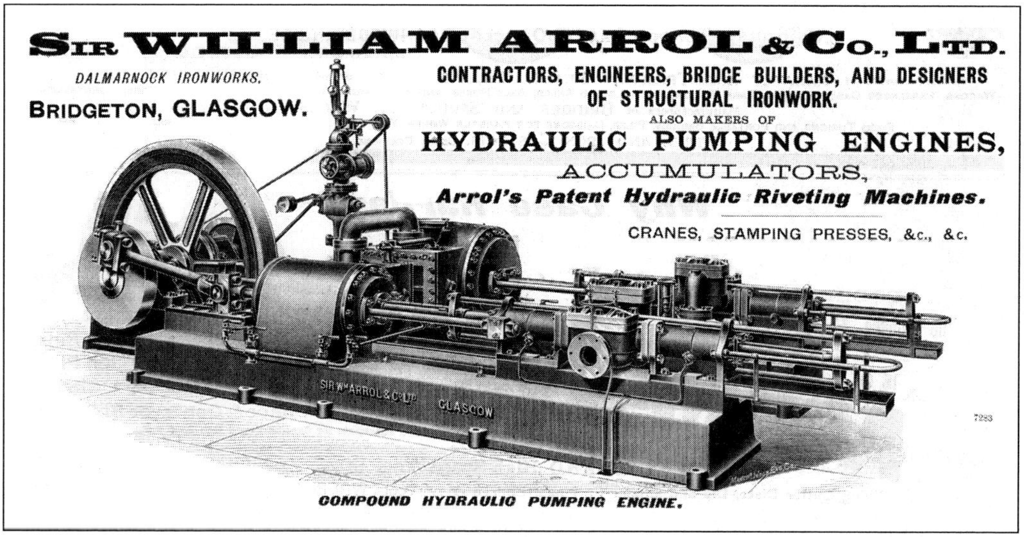

MAK/018 - Arrowsmith, Sinclair & Co., 14 Upper William Street, Liverpool.

1) Advertisements:
1903.

Reverse: 1903 The Klein Engineering Co., 94 Market Street, Manchester, - condensing and cooling plants.

MAK/019 - Ashton, Frost & Co., Bank Top Foundry, Blackburn.

2) Letters:
MS Letter to GW from F. A. Howarth, dated 21 June 1958. He started in the works in 1896 and remained there until 1926.

2 undated letters, but probably 1958 from Mrs M. Pemberton, whose relatives worked at Ashtons.

From Daisyfield flour mill, Blackburn, 17 March 1960, confirming they had an Ashton Frost engine which was broken up some years before.

From West Grove Mfg Co., Ltd, Allanby Mill, Westhoughton, Lancs., 19 January 1981, confirming they had just scrapped their Ashton Frost engine.

3) MS material:
3 sheets of MS notes by George Watkins.

4) Photographs:
Print of 2 triple expansion pumping sets. Elksley pumping station, Lincoln WW. 'Started 9/8/1911' on engine. (see below).

5) Printed material:
500 HP compound mill engine, made for Codling & Hodgkinson, Mill Hill Mill, Blackburn, ex *Mechanical World* 18 August, 1898, pp 66-67. (see p.24).

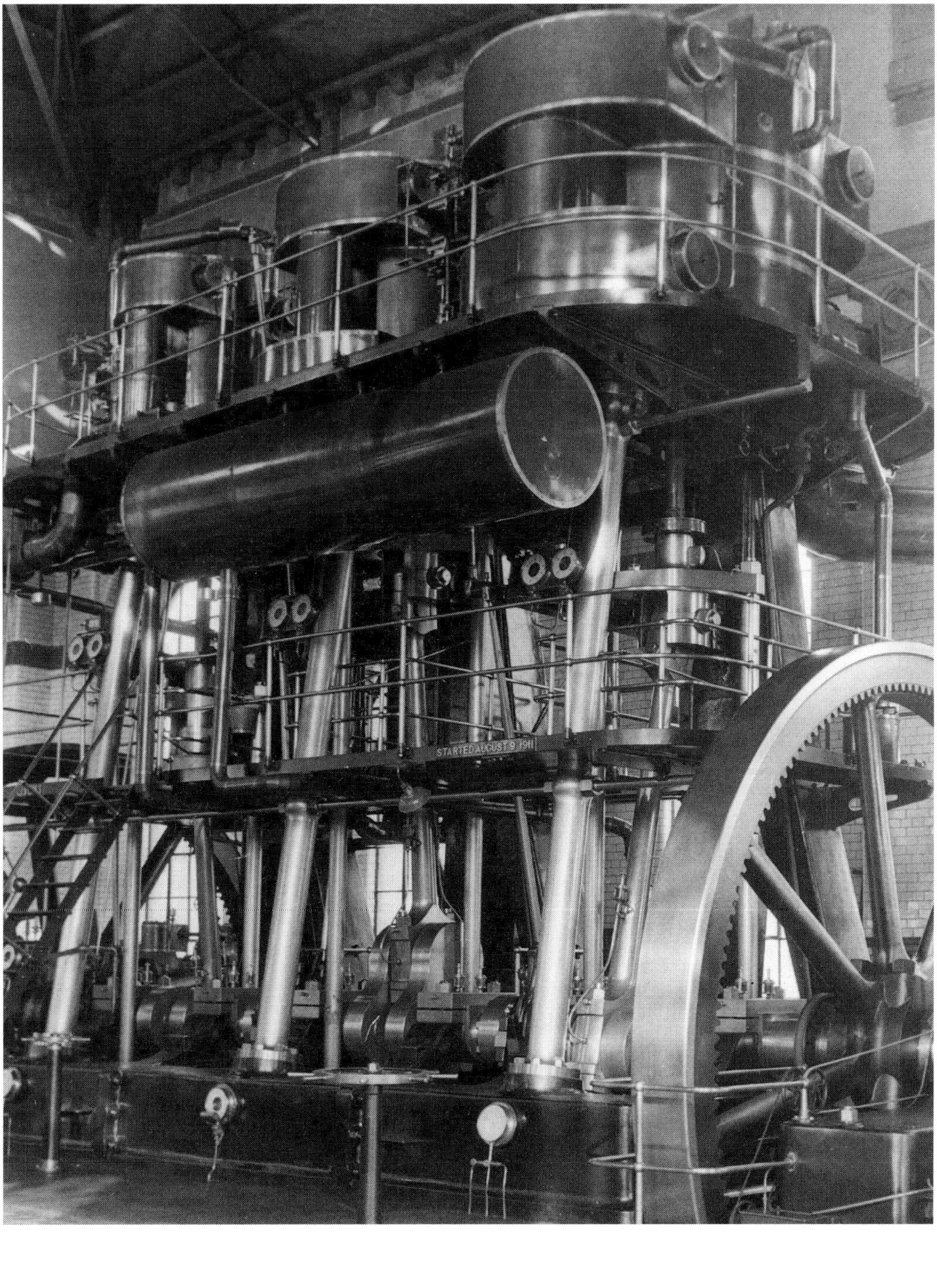

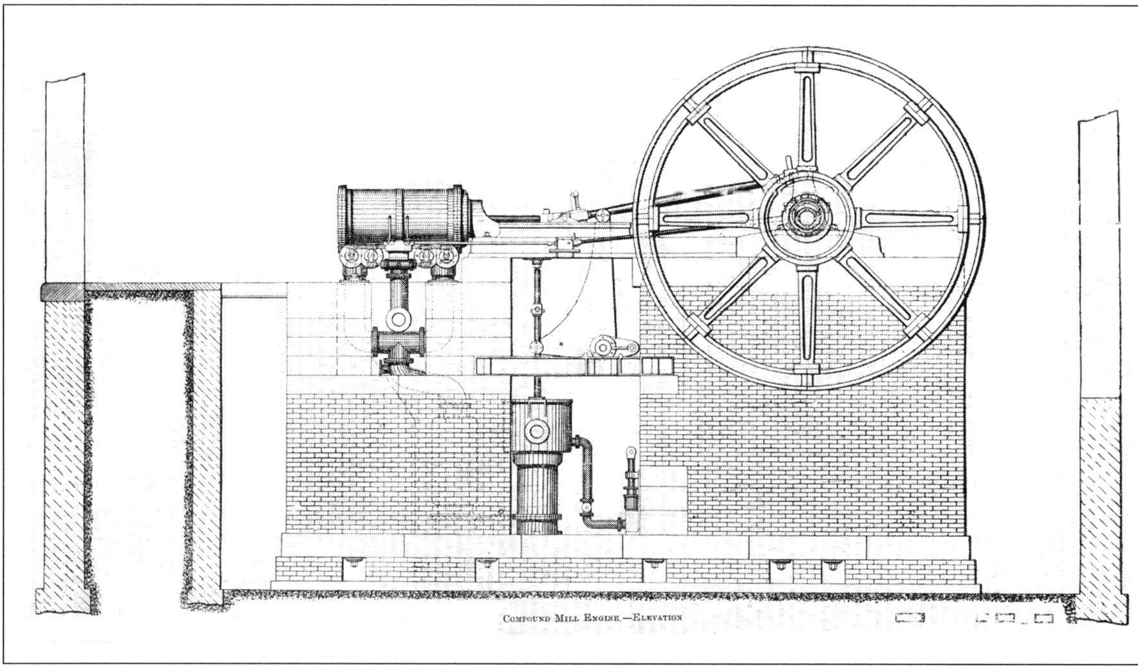

Compound Mill Engine.—Elevation

MAK/020 - Ashworth & Parker, Bury, Lancs.

2) Letters:
From Ashworth & Parker, 30 April 1963 about their engines.

6) Trade Material:
Leaflet: E. 59 about the engines they manufactured, 4pp, c 1960.

MAK/021 - Atlantic Works, Broadheath, Altrincham, Cheshire.

1) Advertisements:
6 March, 1896 regarding auction of the works.

MAK/022 - Atlas Engine Company, Atlas Works, Oozells Street, Birmingham.

1) Advertisements:
1879. Mentions exhibiting at Paris Exhibition 1878, and Royal Show 1879. (see p.25).

Reverse: 1897 Thomson & Sterne, Crown Iron Works, Glasgow; Clayton, Howlett and Venables, Atlas Works, Harrow Road, London.

Royal Agricultural Show at Kilburn, Stand 434.

THE ATLAS ENGINE COMPANY, LIMITED,

ATLAS WORKS, OOZELLS STREET,

BIRMINGHAM.

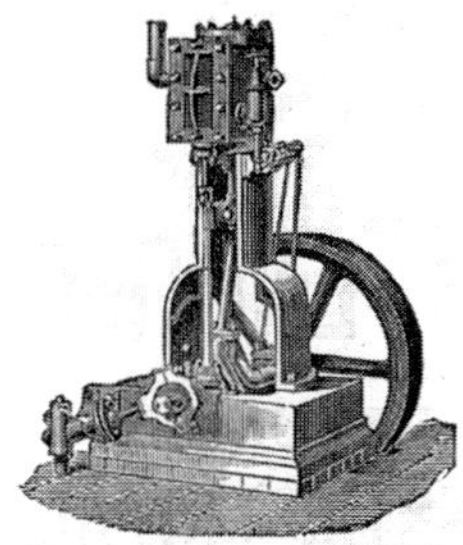

THE "ATLAS" VERTICAL ENGINE.

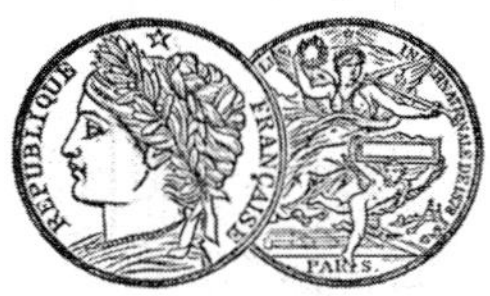

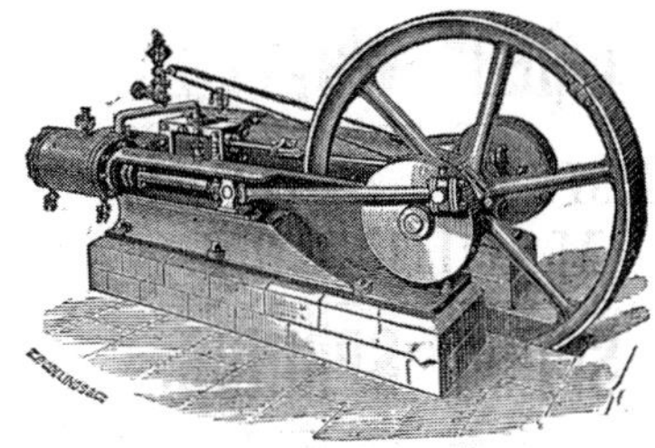

THE "ATLAS" HORIZONTAL ENGINE (Coupled)

TWO SILVER MEDALS,

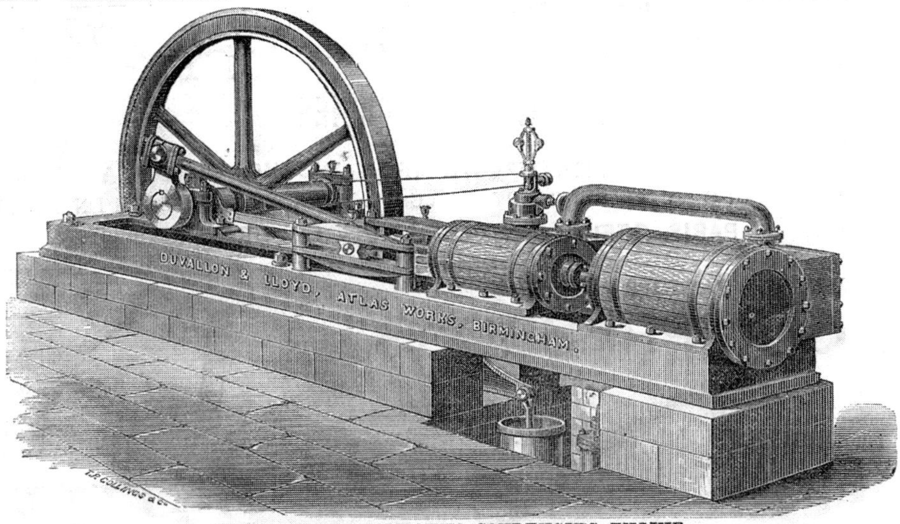

THE "ATLAS" COMPOUND CONDENSING ENGINE.

PARIS, 1878.

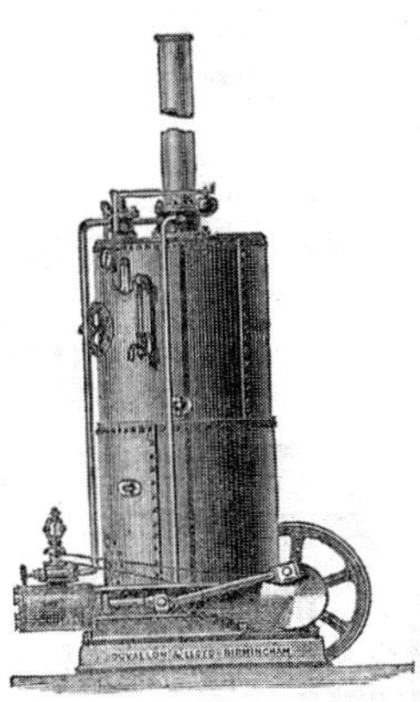

THE "ATLAS" HORIZONTAL ENGINE & VERTICAL BOILER.

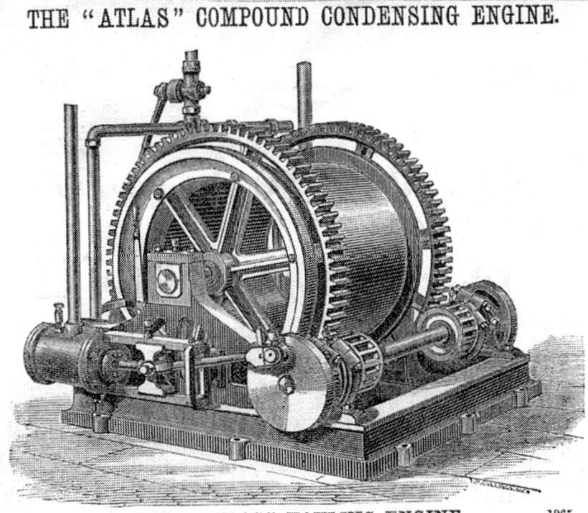

THE "ATLAS" HAULING ENGINE.

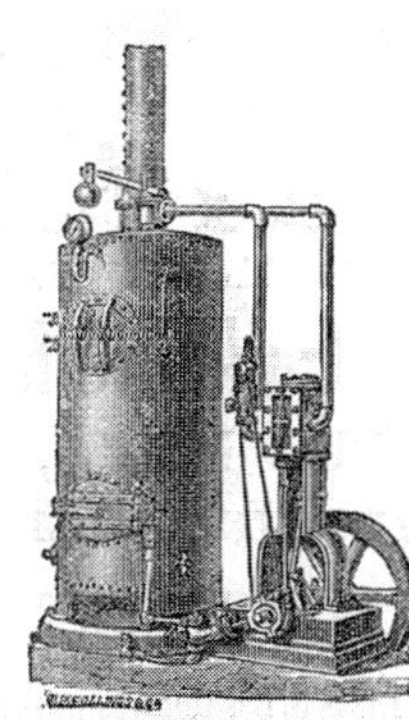

THE "ATLAS" VERTICAL ENGINE AND BOILER.

1965

LONDON OFFICE, 4, Broad Street Buildings, Liverpool Street, E. C.

MAK/023 - Babcock & Wilcox, Co., New York and Glasgow.

5) Printed material:
Photocopy of booklet: *The Babcock & Wilcox patent stationary steam engine*, New York 1870, 32 pp.

Printed catalogue: *Steam*, 8th British edition, 1913, 211 pp. Contains many fine photographs of boiler installations at named sites. Photographs shown are from this catalogue (pp.12 & 24).

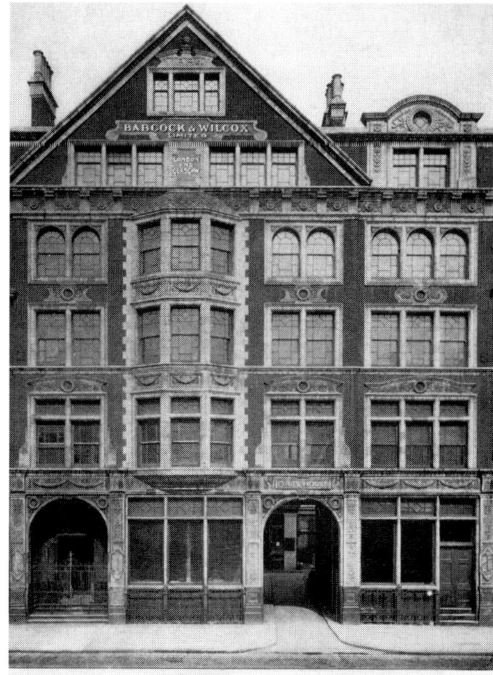

BABCOCK & WILCOX LTD., HEAD OFFICES, ORIEL HOUSE, FARRINGDON STREET, LONDON.

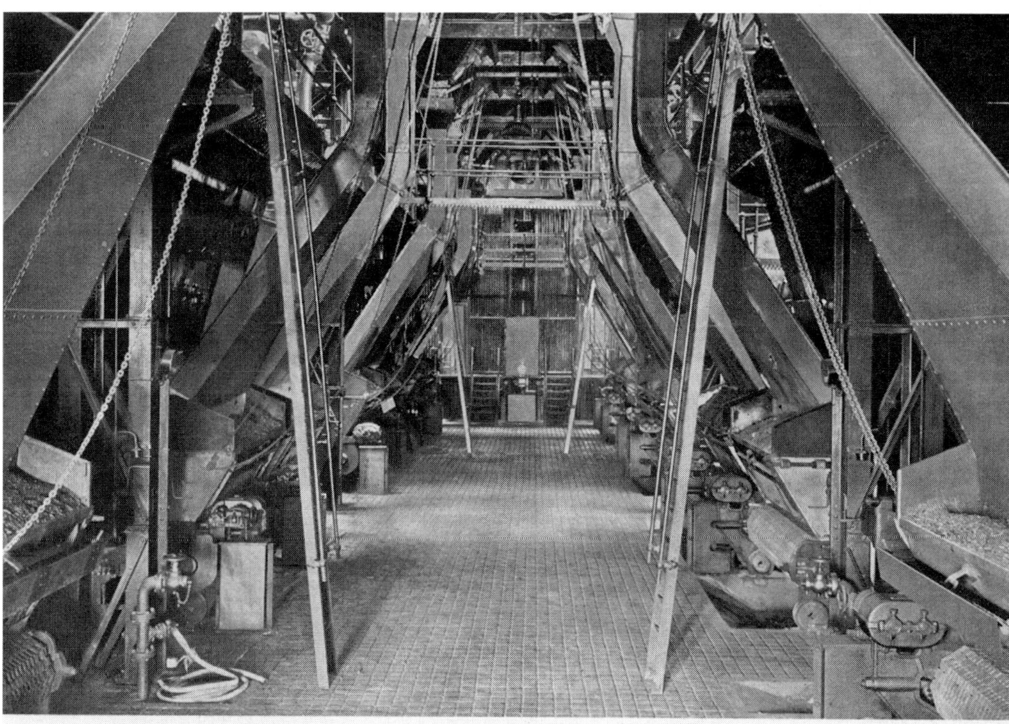

NEWCASTLE-ON-TYNE ELECTRIC SUPPLY CO., LTD., DUNSTON POWER STATION.
Eight Babcock & Wilcox Boilers (Marine-type), each of 6,725 square feet heating surface, fitted with Superheaters and Chain Grate Stokers

MAK/024 - Bagnall, Stafford.

1) Advertisements:
1878, 1879 (see bottom image), 1880 (see middle image).

2) Letters.
Letter to GW from Bagnalls, dated 19 November 1953. acknowledging an earlier one of his. Letterhead shown below.

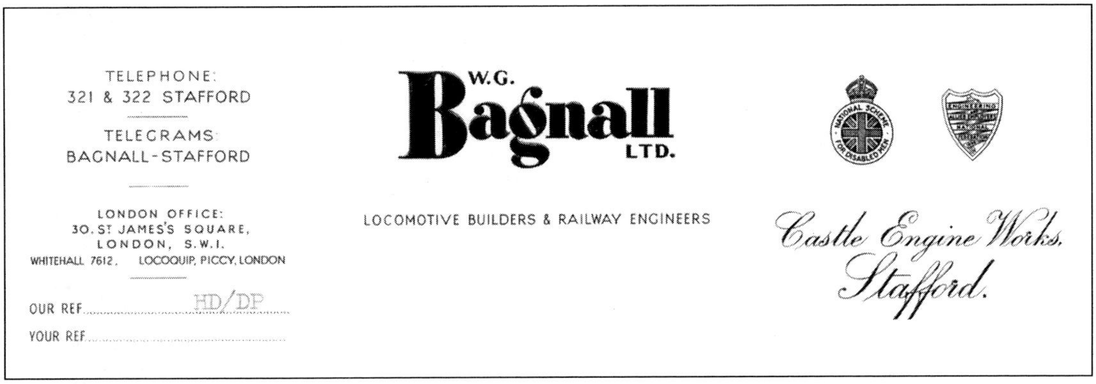

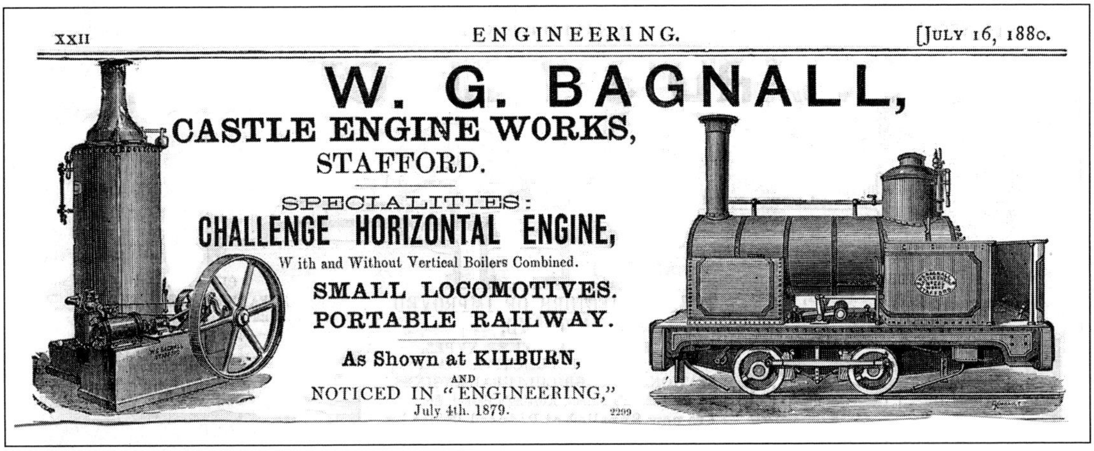

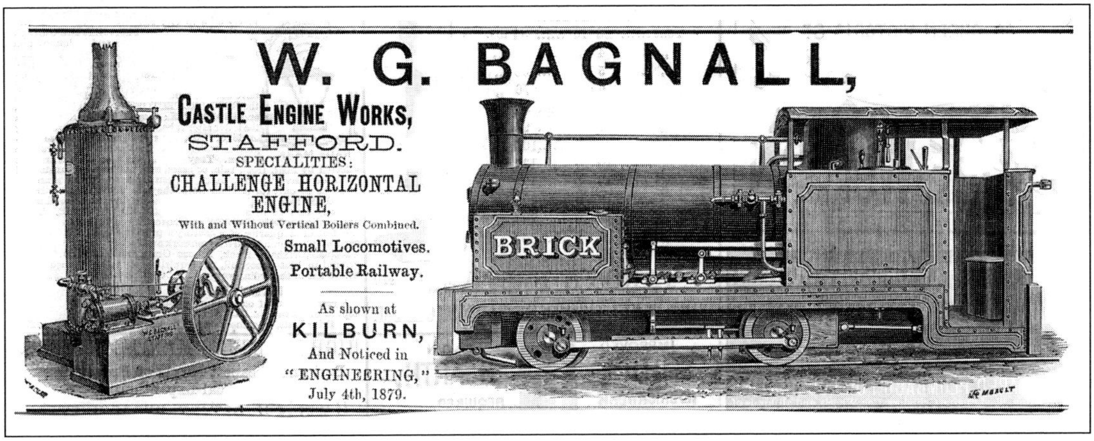

MAK/025 - J. Bagshaw & Sons, Victoria Foundry, Batley, Yorks.

5) Printed material:

Copy of plate of compound mill engine from *The Engineer*, 5 November 1875. Lacks text. (see below).

Brief article describing Bagshaw's method of compounding, from *Engineering*, 11th March 1887.

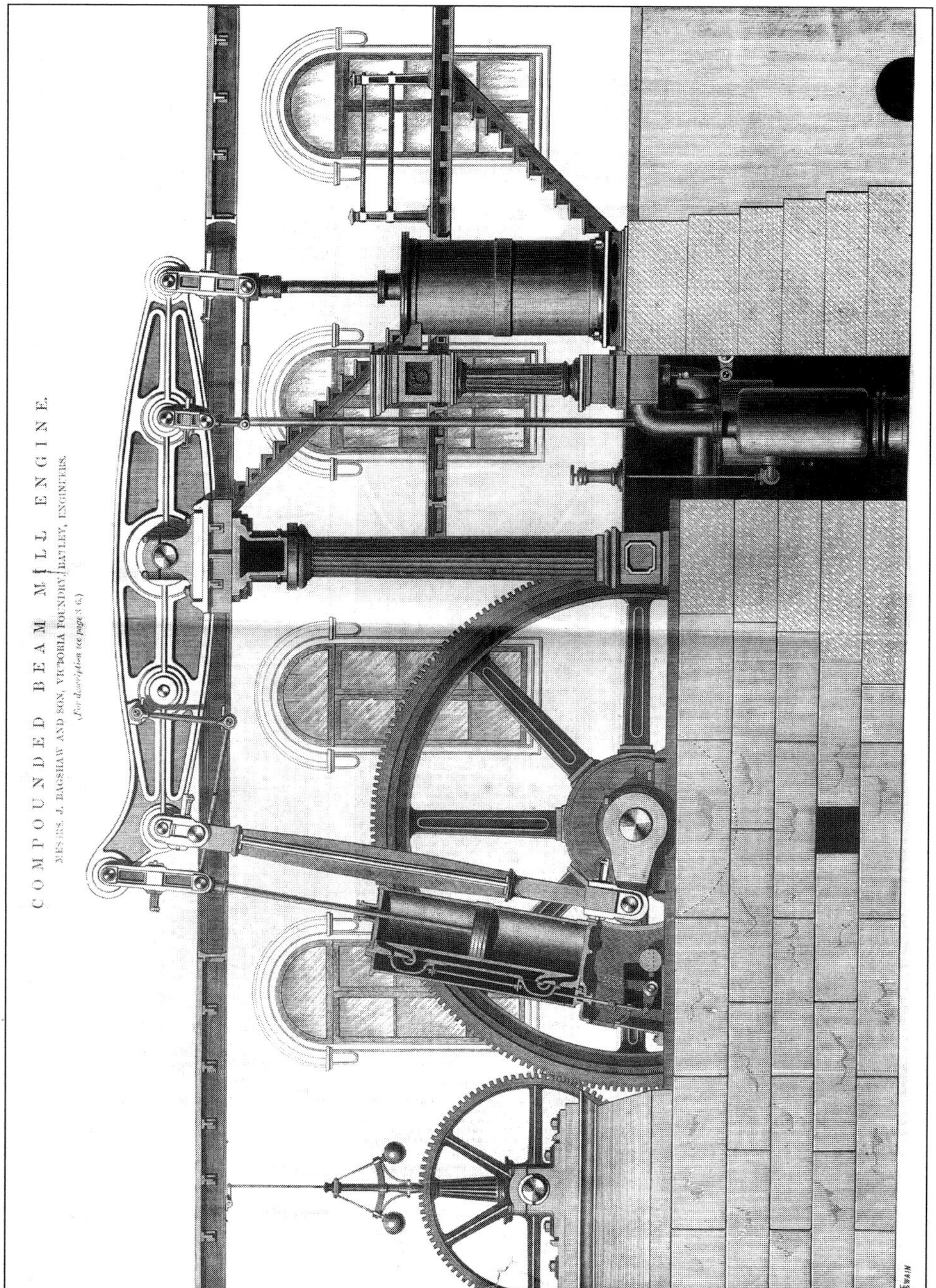

MAK/026 - W. H. Bailey & Co., Albion Works, Salford, Manchester.

1) Advertisements:
1878, 1896.
Reverse: 1896 Cowans Sheldon of Carlisle, - cranes; Hill Clarke & Co., Boston, Mass. USA, - lathes.

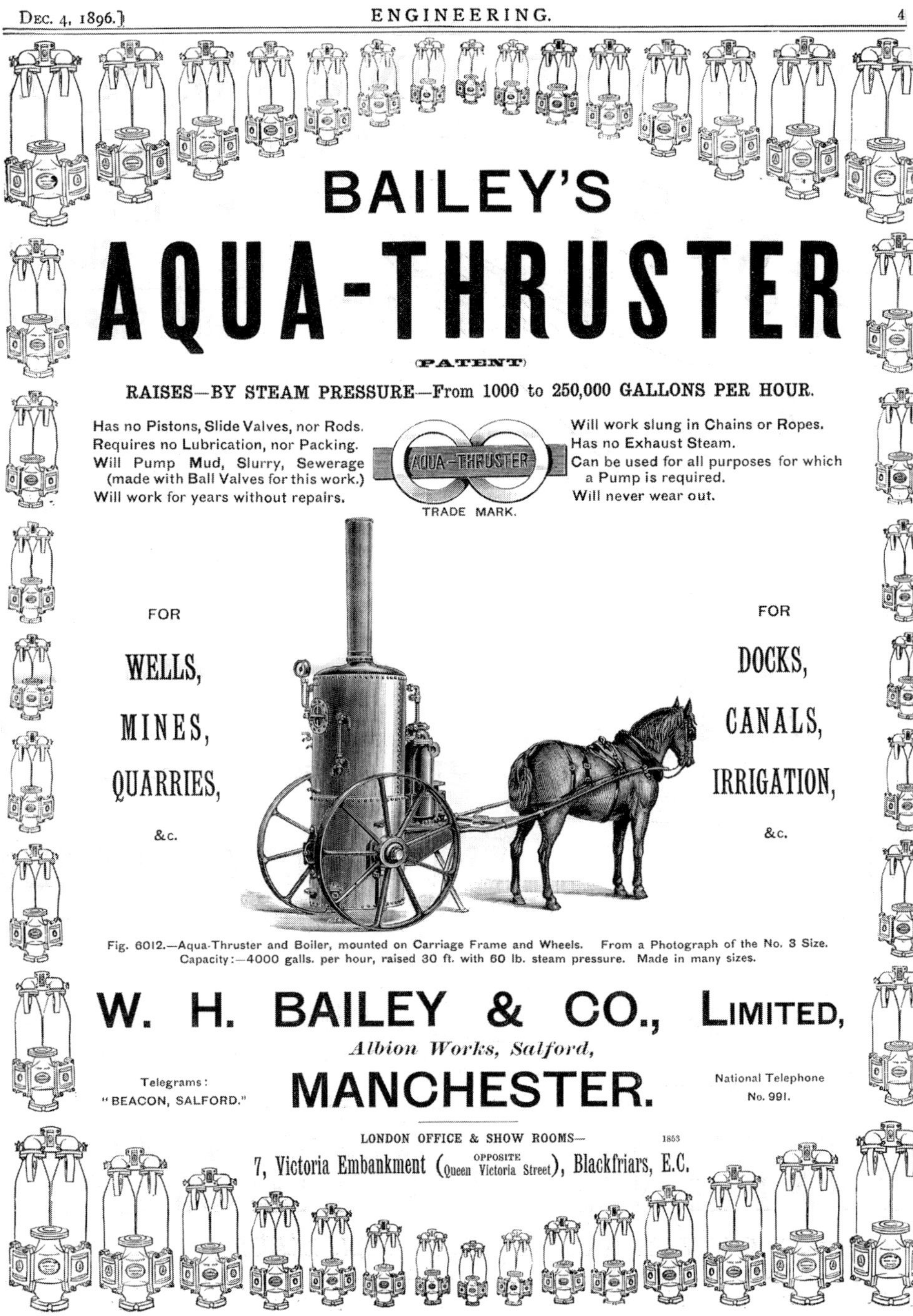

MAK/027 - Barrows & Stewart, Engineers, Cherwell Engine Works, Banbury, Oxon.

1) Advertisments:
1896 auction sale of works.

MAK/028 - Barry, Henry & Co., Aberdeen.

1) Advertisements:
1896.

MAK/029 - Timothy Bates, (Late Pollit & Wigzell), Sowerby Bridge, Yorks.

5) Printed material:
Photocopy from *Mechanical World*, 4 October 1890, with plan of unidentified compound mill engine.
Badly faded copy of page from book about Yorkshire industry, c 1890, Too bad to read.

MAK/030 - Walter Bates Co., Inc., Joilet, Illinois, USA.

2) Letters:
Typed letter to GW from Walter Bates, dated 20 September 1954 about the Bates Corliss engine at World's Fair, 1903.

MAK/031 - Bedford Engineering Co., Bedford.

1) Advertisements:
1903.
Reverse: 1903 Wm. Buckley, Patent Piston Works, Sheffield.

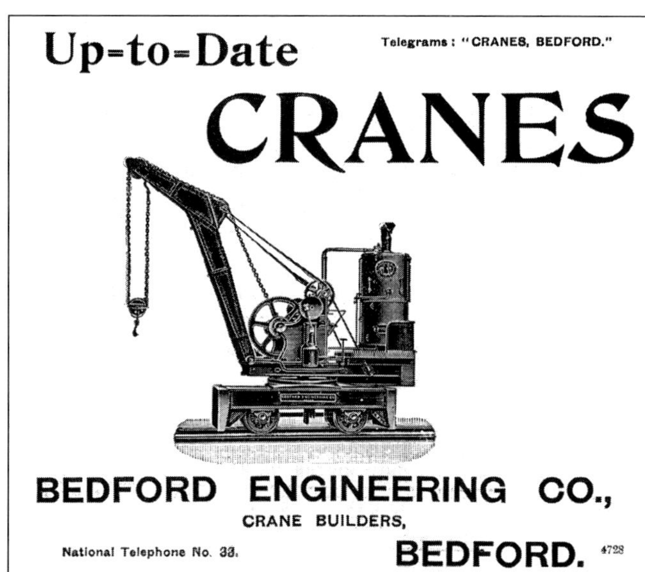

MAK/032 - Thomas Beeley & Son Ltd, Hyde, Manchester.

1) Advertisements:
1912.
Reverse: 1912 Vaughan & Sons, Ltd, Manchester – overhead travelling cranes.

MAK/033 - Beitrix & Co., St Etienne, France.

5) Printed material:
Article from *Engineering,* 9 November 1900 describing 185 kW horizontal engine at the Paris Exhibition.

MAK/034 - Belliss & Morcom, Birmingham.

1) Advertisements:
1894, 1895, 1896, n.d. (c1900), 1903, 1908.

2) Letters:
1) Letter to GW from Belliss & Morcom, dated 11th August 1966 relating to the list of rope drive installations and the set of drawings noted in 7) below.

4) Photographs:
Bad prints of 3 Belliss engines Sellotaped together, fading.
Print from catalogue of high speed engine with rope drive from Sparkthwaite Mills.

5) Printed material:
Photocopy of article about Belliss and Morcom by C. E. Page in *Model Engineer*, 13th December 1956, pp 823-833 illus.

6) Trade literature:
Catalogue: *Standard Manufactures* ?1941, 16pp.
Catalogue: *Textile mill driving pamphlet* 1928 ed. 12pp. Covers split.

7) Other material:
TS list of various rope drive installations by Belliss & Morcom, 2pp. n.d. but 1966, see 2) above.

Collection of blueprints of engines:
8369 Engine at Bowling Mill Combing Co., 1933.
Turbo geared generator and rope drive for Tomkinsons, 1934.
9321 pulley engine for Messrs Taylor Tunnicliffe, 1937.
10029 pulley engine for Wm. Paton, Ltd, 1942.
250 bhp set at Brookfield Spinning Co., n.d.
150 bhp set at Brewer's food supply, 1951.

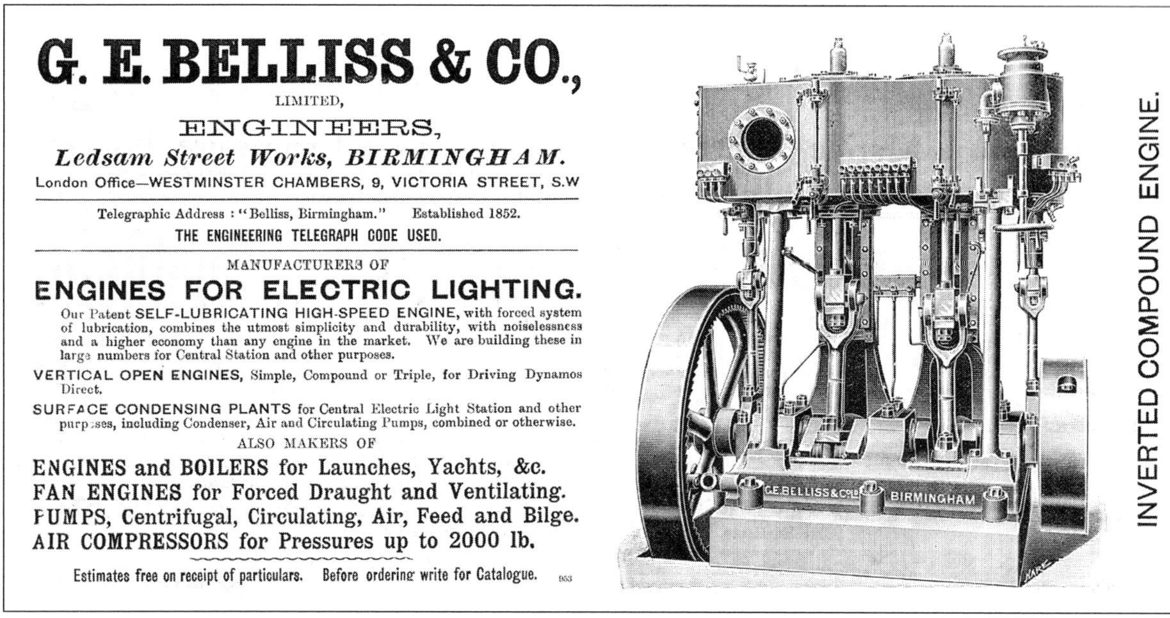

Advert dated 1896.

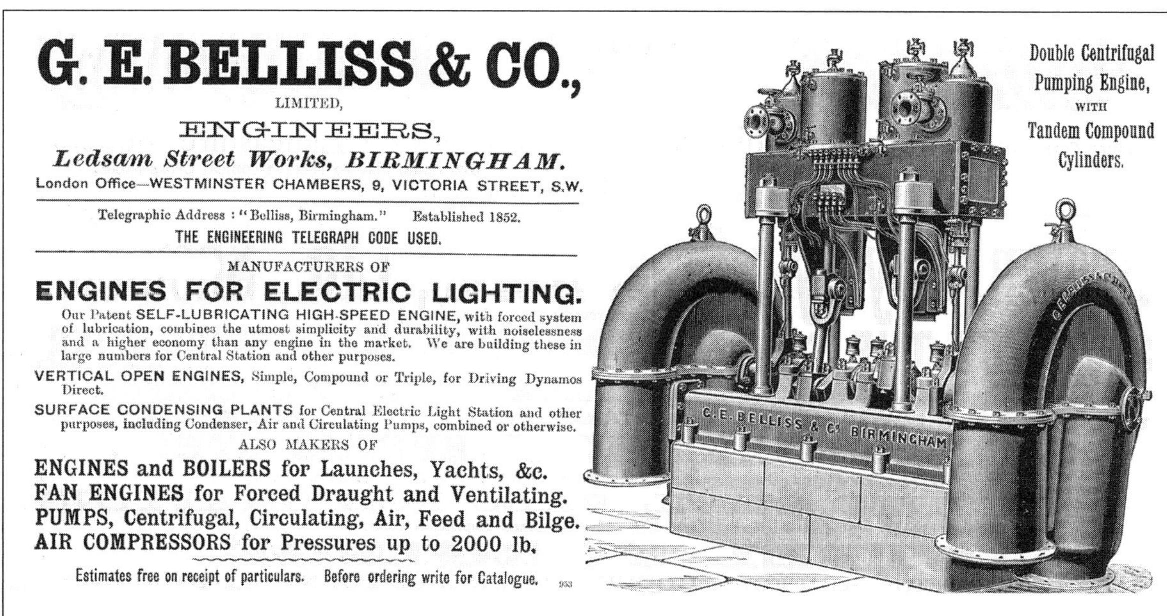

Advert c 1900.

MAK/035 - Bever & Dorling (late W. & J. Cardwell), Dewsbury Foundry, Dewsbury.

1) Advertisements:

1888 1889 c 1890. 1894, 1887 (2), 1903, 1905 (3), 1906, 1909, 1912

Reverse: 1887 Ruston Proctor & Co.,; Glenfield & Kennedy, Kilmarnock.

Reverse: 1905 Clyde Structural Iron Co. Ltd.

Reverse: 1909 W. G. Bagnall, Locomotives.

4) Photographs:
Collection of photocopies of photographs. Origin unknown.
Single cylinder horizontal engine.
Vertical engine.
Twin horizontal compound pumping engine. W. Maxwell, Engineer, Djokja.
[Bever and Dorling licensee?]
View of Bever and Dorling's workshop at Low Moor.

5) Printed material:
Copy of patent No. 1666 of 1884 for Bever and Dorling's high speed steam engine.

Photocopy of abridgement of patents 1889 with note about patent no 9369 for Bever's high speed compound engine.

Copy of Patent No 20226 of 1911 for Bever and Dorling's overwinding gear.

Copy of printed description of Bever and Dorling's patent balanced high speed engine. Pencil note by GW about Wolverhampton Exhibition 1902 or 1903. 4pp.

Photocopy of part of booklet, published by United Sugar Engineers Ltd illustrating Bever and Dorling's engines, n.d., but c 1910.

6) Trade Literature:
Collection of photocopies of parts from Bever and Dorling's catalogue, n.d. but c 1920.
2 copies of Bever and Dorling's patent winding engine controller, 3pp.
Self contained winding engines 2pp.
Pumping engines 7pp.
Various horizontal engines 8pp.
Various winding engines, 11pp.

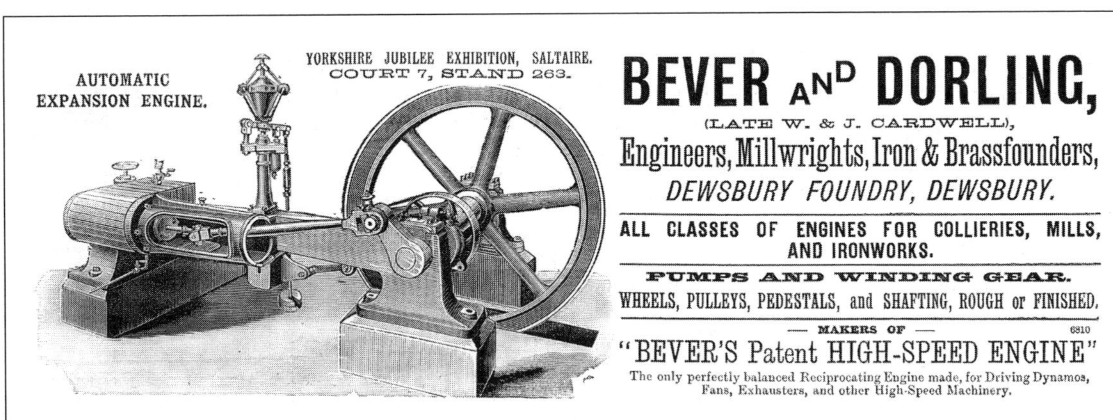

Advert dated 1887.

MAK/036 - Beyer Peacock & Co., Gorton, Manchester.

2) Letters:
Letter to GW dated 20 November 1951 about geared and rack locomotives.

MAK/037 - John R. Bird, 10 Morrison Street, Kingston, Glasgow.

1) Advertisements:
1894.

Same side: 1894 Drysdale & Co., Bon Accord Engine Works, Glasgow. - centrifugal pumps.

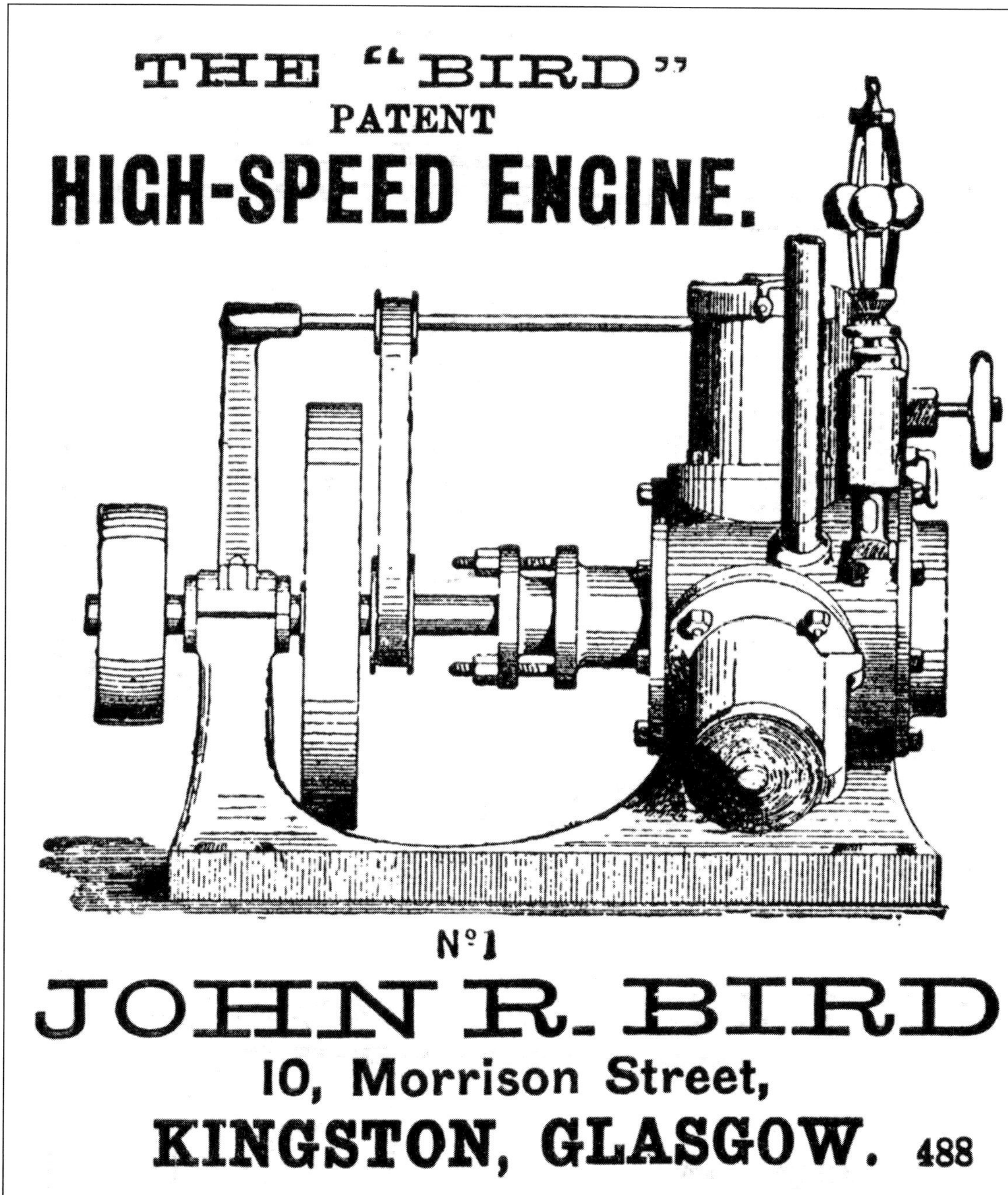

MAK/038 - Blair Engineering Co., Peckham, London, S. E.

1) Advertisements.
1894.

MAK/039 - Blairs Limited, Glasgow Engineering Works, Woodville Street, Govan, Glasgow.

2) Letters:
Letter to GW from Blairs, dated 6 November 1958 about horizontal engine. Relates to drawing and leaflet: noted below.

6) Trade literature:
Page 17 from Blairs' sugar mill catalogue: no date but c 1958.

7) Other material:
Blueprint of engine 3863. Horizontal coupled steam engine, n.d.

MAK/040 - Bollincx & Co., Brussels, Belgium.

5) Printed material:
Engraving from *Engineering* of 1000 hp compound condensing engine at the Paris Exhibition. No text.

MAK/041 - Bolton Iron & Steel Company, Bolton, Lancs.

1) Advertisements:
1905 auction sale of the works.

MAK/042 - Borsig, Berlin, Germany.

5) Printed material:
Article from *Engineering* 10 August 1900 about 2500 HP vertical engine at the Paris Exhibition.

6) Trade literature:
Leaflet about Borsig engines 1850-1950 c 1951, 4pp, with 3 loose sheets of relevance, some in German.
Product list, 4pp n.d. in English.
Product leaflet, 2pp n.d. in German.
Photocopy of leaflet about Borsig engines, 4pp n.d.

MAK/043 - Bowling Iron Co., Bradford, Yorkshire.

1) Advertisements:
1894 (2) These are photocopies on A4 sheets of full page advertisements from *Engineering.*

4) Photographs:
2 - 60HP Woolf compound beam engines, at John Priestman's Co's Ashfield Mills, Bradford. (see below).
Print of above engine photographed by C. H. Wood of Bradford. (see p.37). Catalogue says " courtesy of WH Allen & Co Ltd (engine makers)".

5) Printed material:
2 pages torn from an issue of *Hepolite Bulletin* about the Ashfield Mills engine, illustrated with the above prints. [Hepolite were piston ring manufacturers.] n.d. but c 1960.

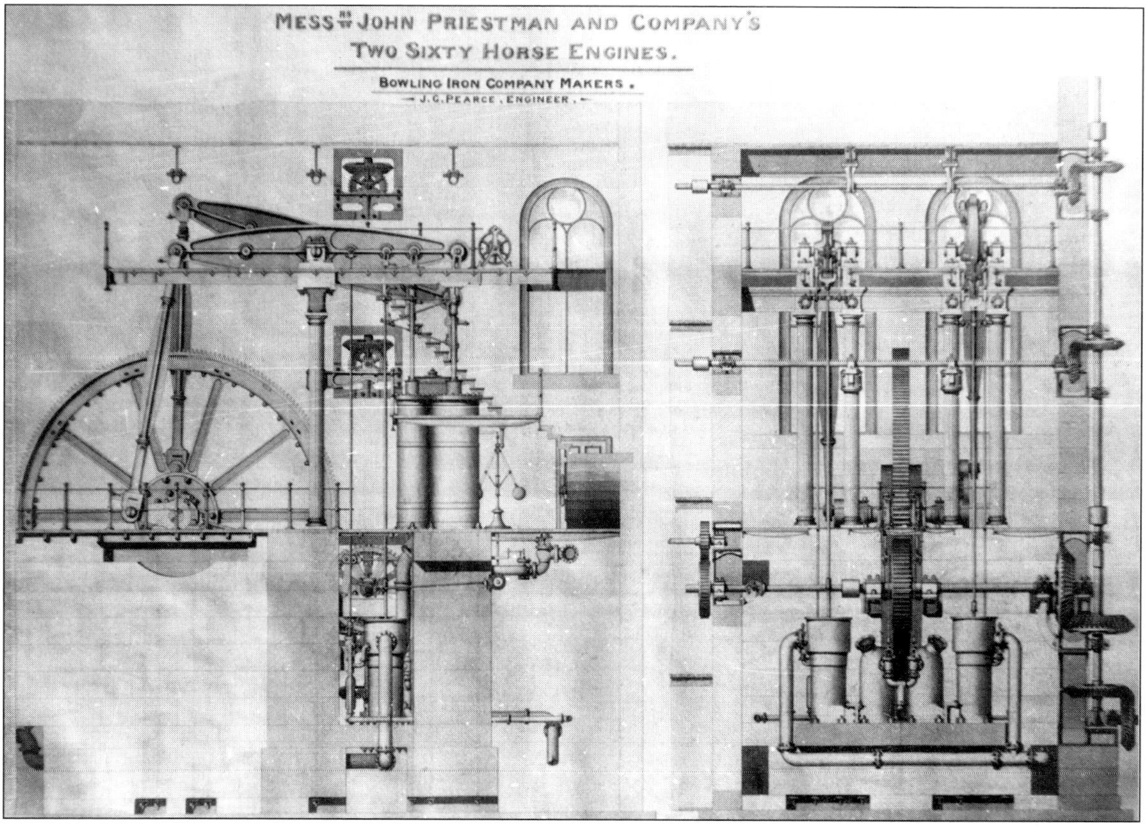

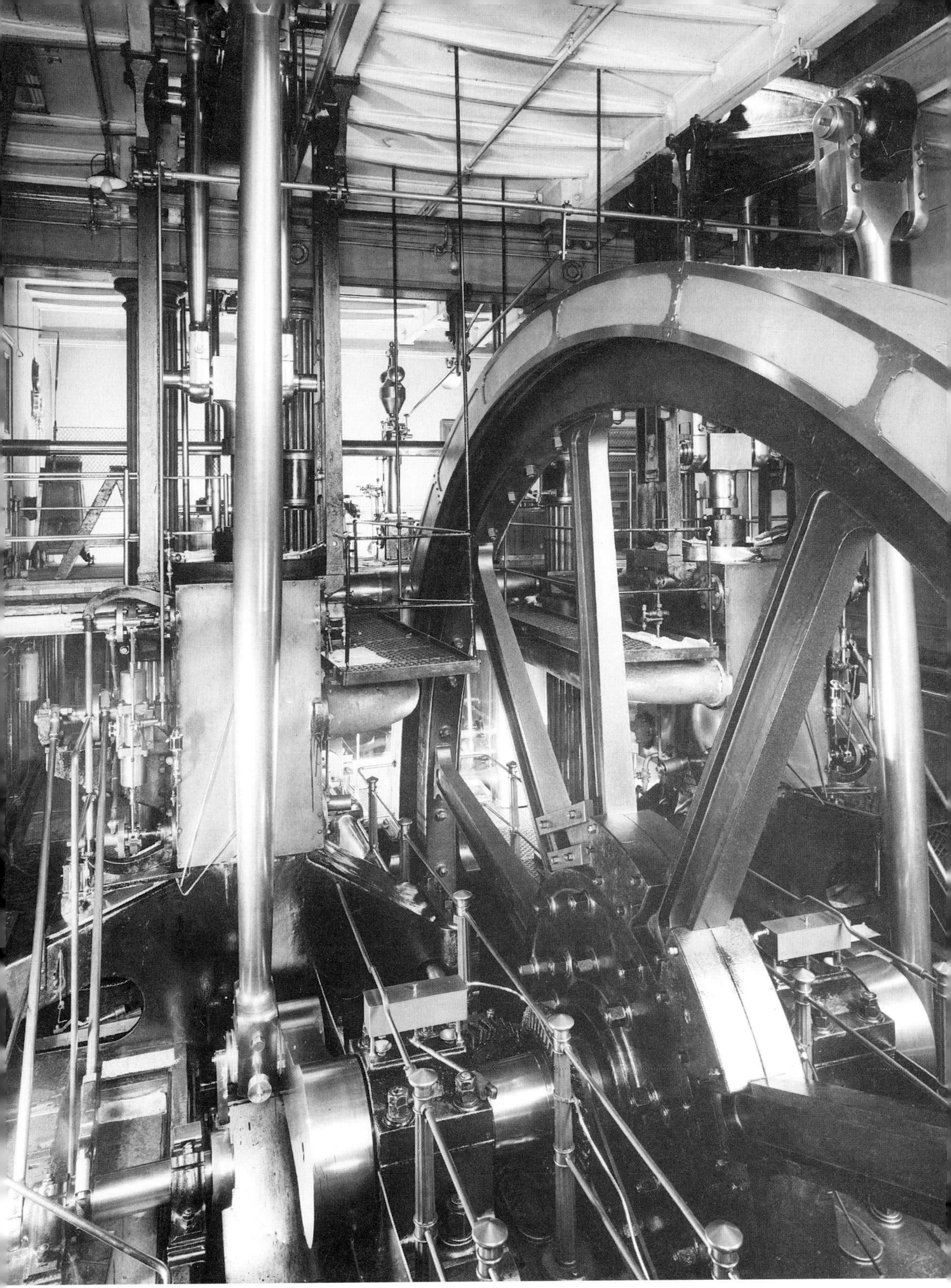

MAK/044 - Bracewell & Pickup, Burnley, Lancs.

5) Printed material:
Offprint [with photocopy] from unidentified periodical n.d. , but c 1880, 14pp. It comprises:

Elevation of a pair of condensing steam engines 500nhp. At mill belonging to J. & J. Craven, Keighley, Yorks. Plate missing.

Bracewell & Pickup's patent compound steam engine. Horizontal compound, illus.

A large example of mill engines. [J. & J. Birley, Burnley].

MAK/045 - Bradley & Craven, Wakefield, Yorks.

2) Letters:
Letter to GW from Bradley & Craven, dated 28 September 1964. Mentions various early engines they made. States that company records were destroyed only 12 months ago as space needed for new machine shop.

Photocopy of letter from R. E. Hooley, Ruston Paxman Diesels Ltd to John Robinson at Science Museum, dated 15 May 1974 about antecedents of Messrs Bradley & Craven who founded the firm.

5) Printed material:
Article from *Engineering*, 25 November 1887, describing Seller's single-crank compound engine, made by Bradley & Craven. Has Corliss valves. Illus.

Photostat of drawing of above. Origin not known. Fading. (see p.39)

Photocopy of page from *The Marine Engineer*, 1 March 1892, describing a vertical marine engine made by them.

6) Trade literature:
Catalogue: *Coal Mining Machinery* n.d. but c 1960, 60pp.
Catalogue: *Colliery haulage and winding engines* n.d. but c 1930, 78pp. (see below)
Leaflet: *Ceramic Machinery Manufacturers*, 1964, 8pp. Describes the rebuilt works. See letter above.

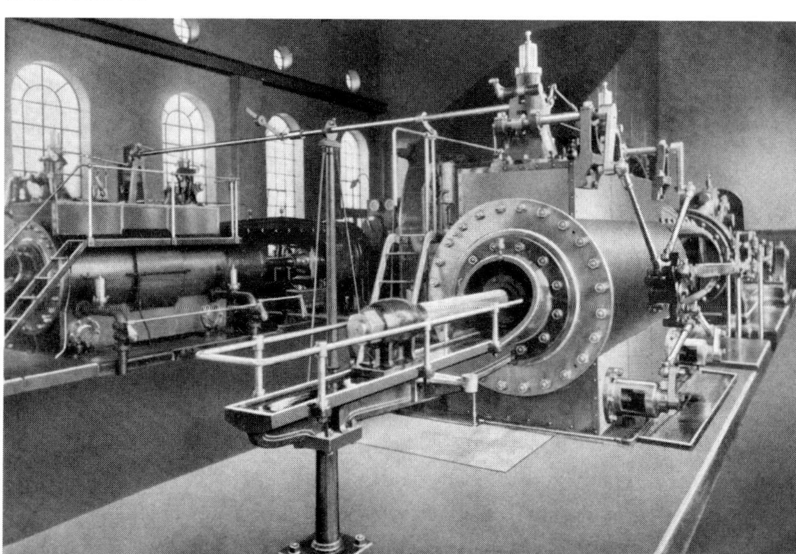

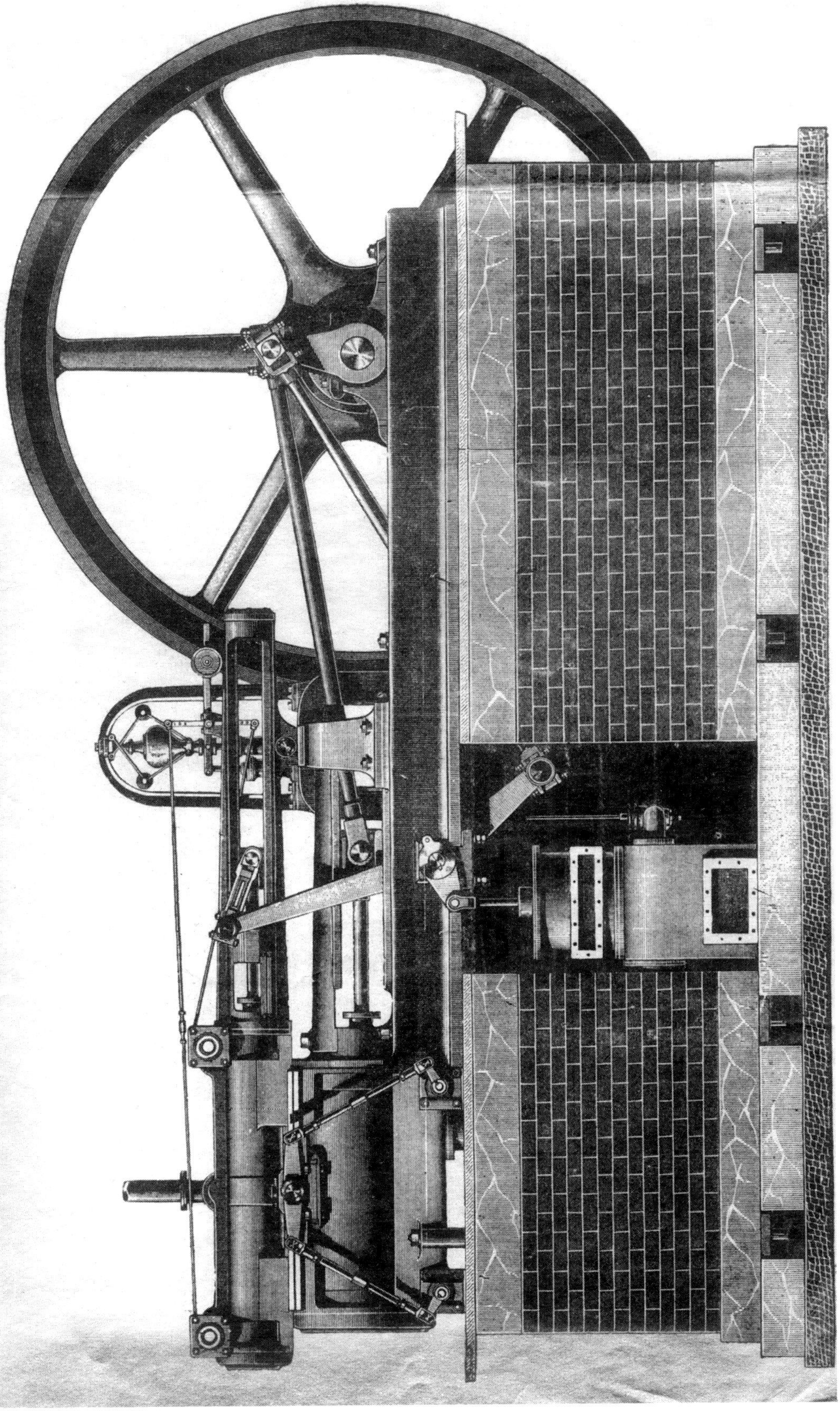

MAK/046 - Brazil, Holborow & Straker, Bristol.

1) Advertisements:
1903; 1907.

6) Trade literature:
Catalogue of the engine and pump department, 1905, 44pp. (see below & p.41).

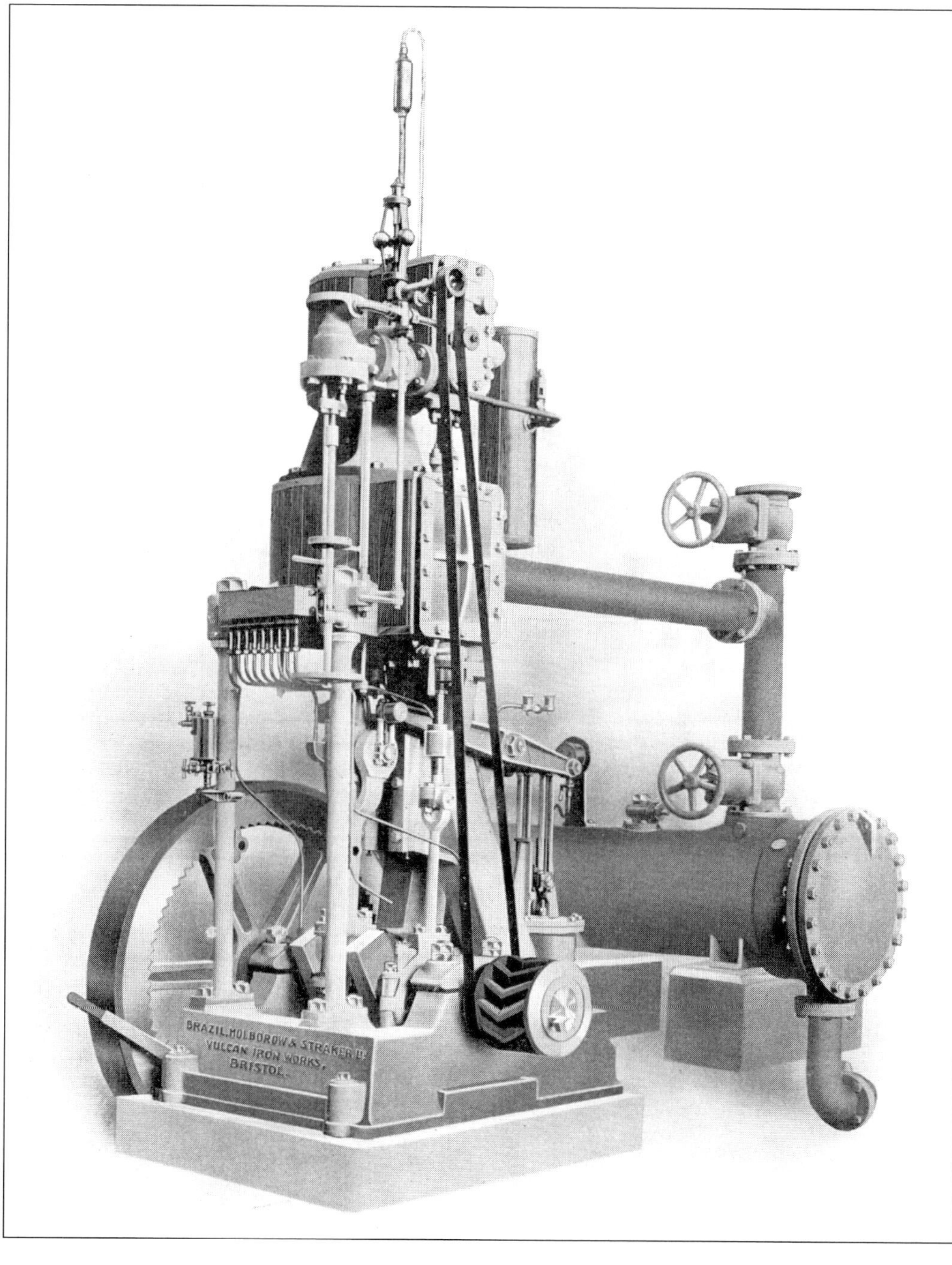

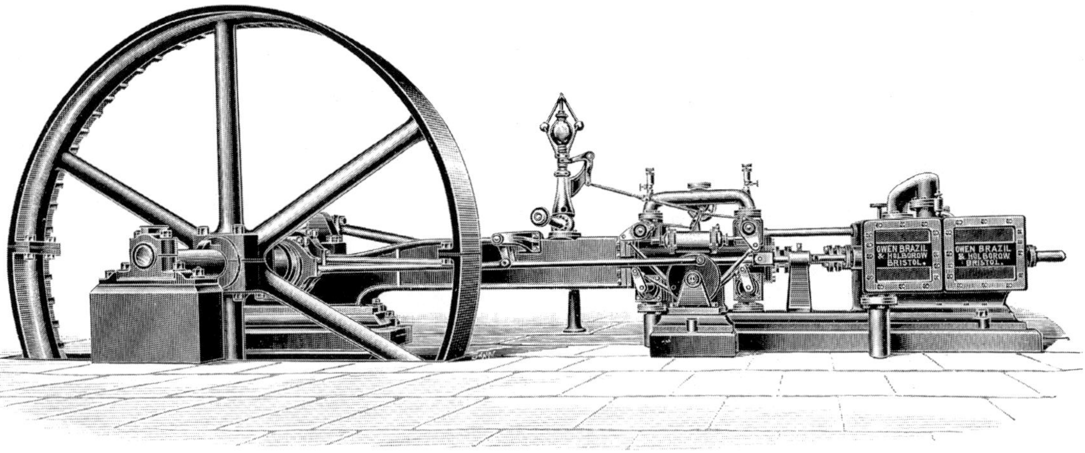

Improved Tandem Compound Non-Condensing Steam Engine, with Corliss Valve Gear

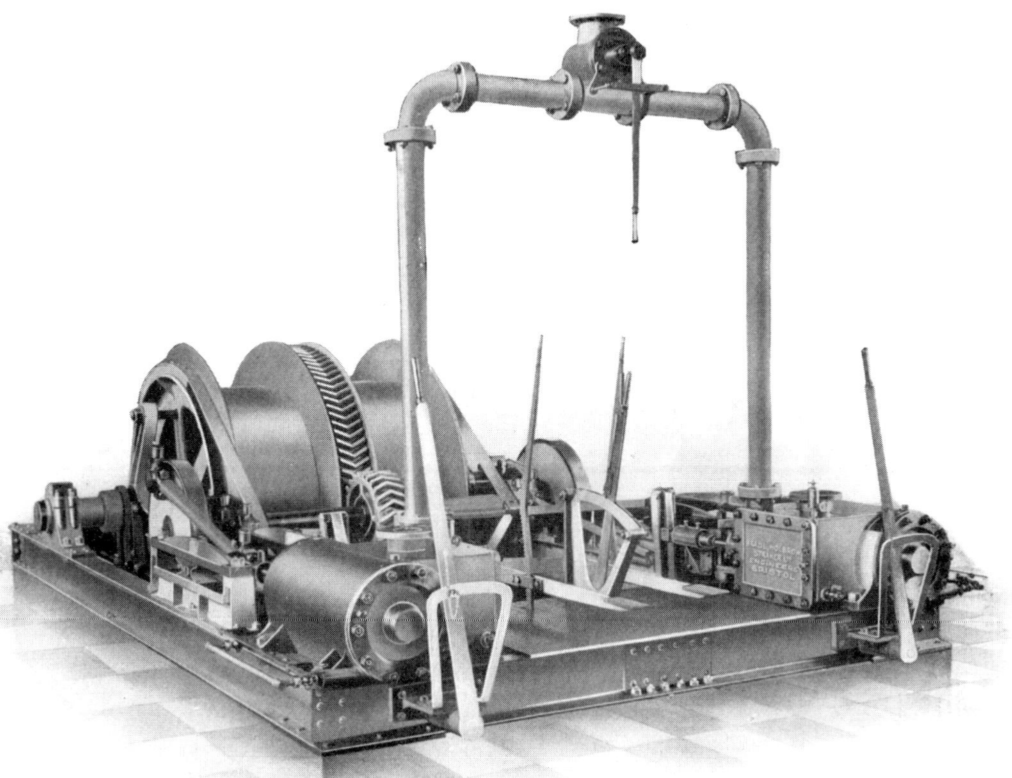

Colliery Winding and Hauling Engine.

MAK/047 - British Electric Plant Co., Ltd, Alloa.

1) Advertisements:
1905.

MAK/048 - British Steel Piling Co. Ltd, Zenith Works, Claydon, Ipswich, Suffolk.

6) Trade literature:
Catalogue of Zenith friction winches, n.d. but c 1966, 12pp. Has illustration of a steam winch.

MAK/049 - Briton Ferry Iron Works, Glamorgan.

1) Advertisements:
1887 for sale of works by auction.

MAK/050 - Thomas Broadbent & Sons. Ltd, Central Ironworks, Huddersfield.

1) Advertisements:
1896, 1930
Same side: Bever and Dorling, Bradford, - double drum electric winder.

2) Letters:
Letter to GW from Broadbents dated 20 September 1961 about the Saddleworth Woollen Co.'s steam engine *Alice*, installed 1899. Mentions a forthcoming centenary publication and photographic prints of *Alice* noted below.

Letter to GW from G. Lumb, 10 Upper Hurst, The Mount Huddersfield. 1 November 1961, a draughtsman at Broadbents. Mentions being responsible for destroying the drawings of engines as they were scrapped. Includes list of engines made 1892 - 1906 and leaflet: noted in 6 below.

Letter from Northern Mill Engine Society, and signed Trevor [? T. Lees, secretary], undated. Mentions Stotts, qv. and has list of engines made by Broadbents.

4) Photographs:
4 - 10 in x 8 in b&w prints of *Alice* noted in 2-1 above. Photographs by Montgomery, 19 Exchange Bldgs, Market Street, Huddersfield.
(see p.44-47).

6) Trade literature:
Single leaf from List E1- *Sugar Machinery Section*. No date but c 1920 by typography. Deals with high speed vertical steam engine designed for driving centrifugal filters for sugar mills. (see p.43).

High-Speed

Vertical Steam Engine

For Use with Types 15 and 20 Centrifugals.

Our Steam Engines can be relied upon to represent the most modern practice, but on account of the designs being periodically improved, we do not bind ourselves to supply the engine exactly as illustration.

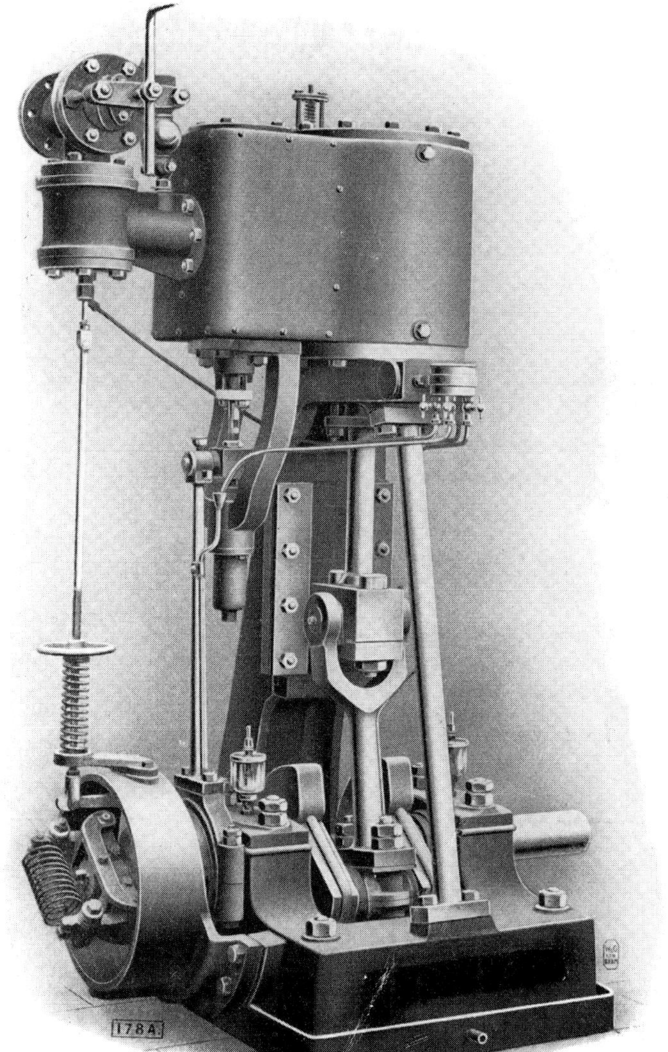

The direction of Rotation must always be outwards.

Thomas Broadbent & Sons, Ltd.,

HUDDERSFIELD, ENGLAND.

Telephone :
1581 (4 lines).

Telegrams :
" Broadbent, Huddersfield."

Established 1864.

Cablegrams :
" Broadbent, Huddersfield."

Codes :
A.B.C. (5th and 6th Editions) :
Marconi ; Bentley's.

REPRESENTATIVES IN ALL PARTS OF THE WORLD.

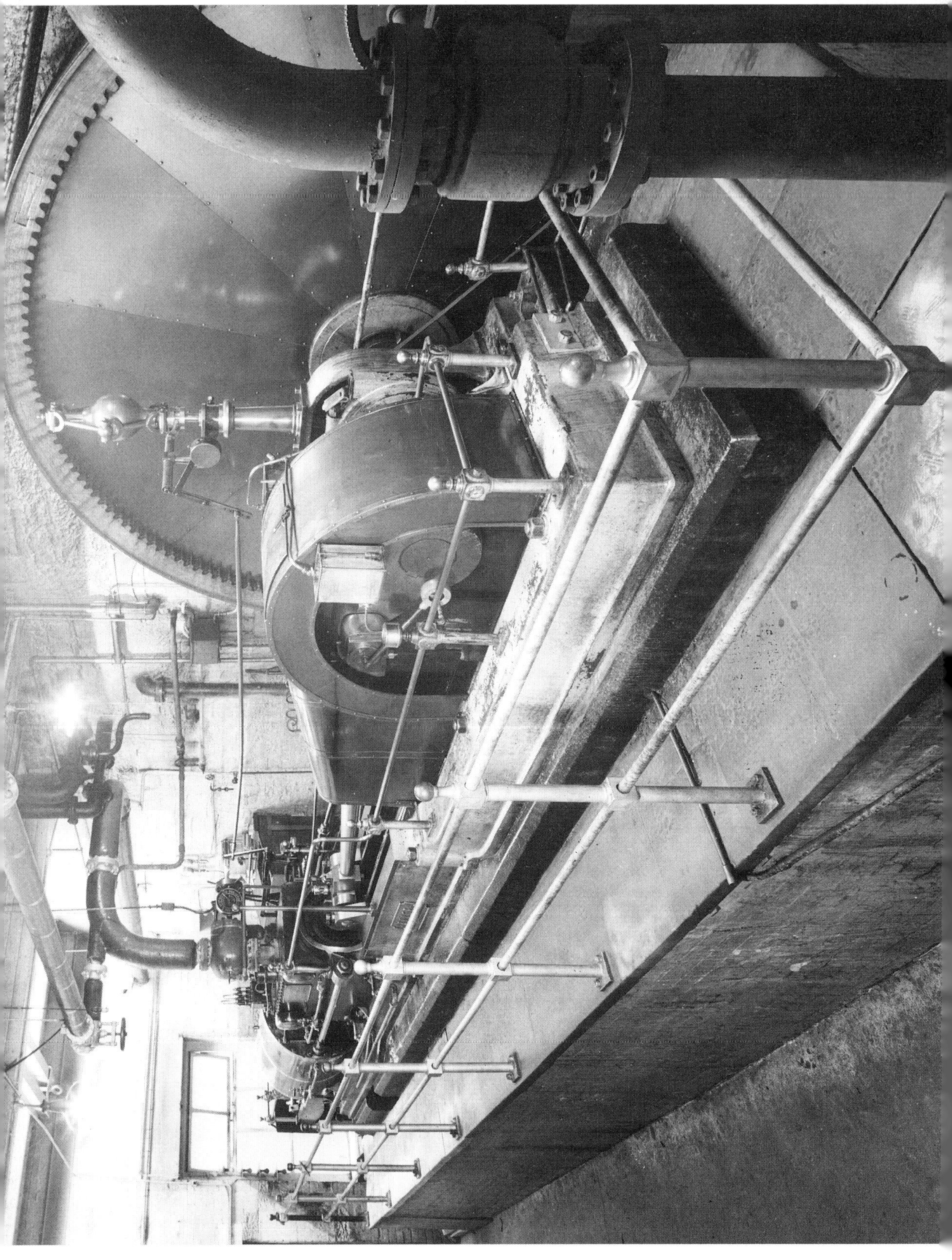

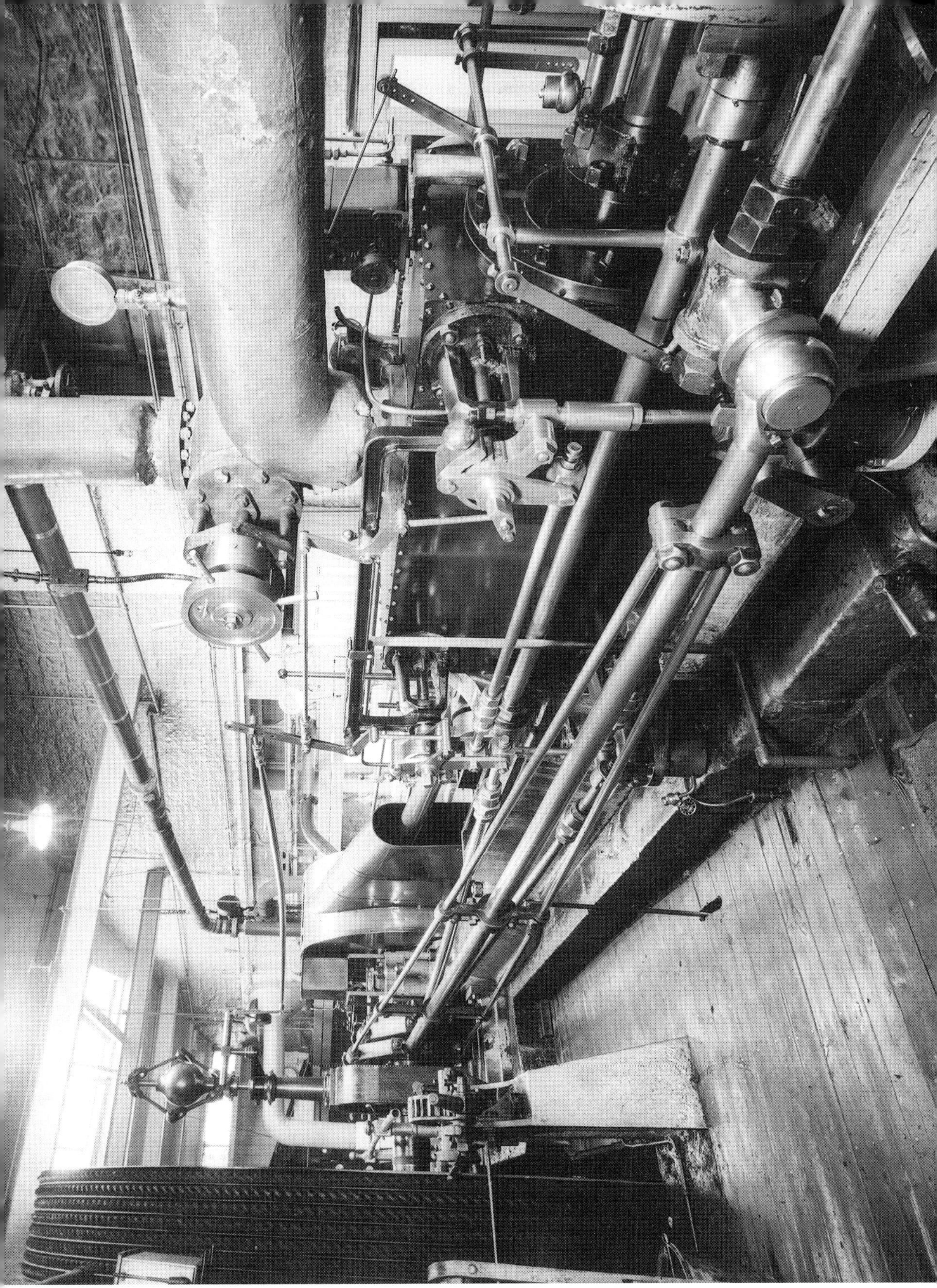

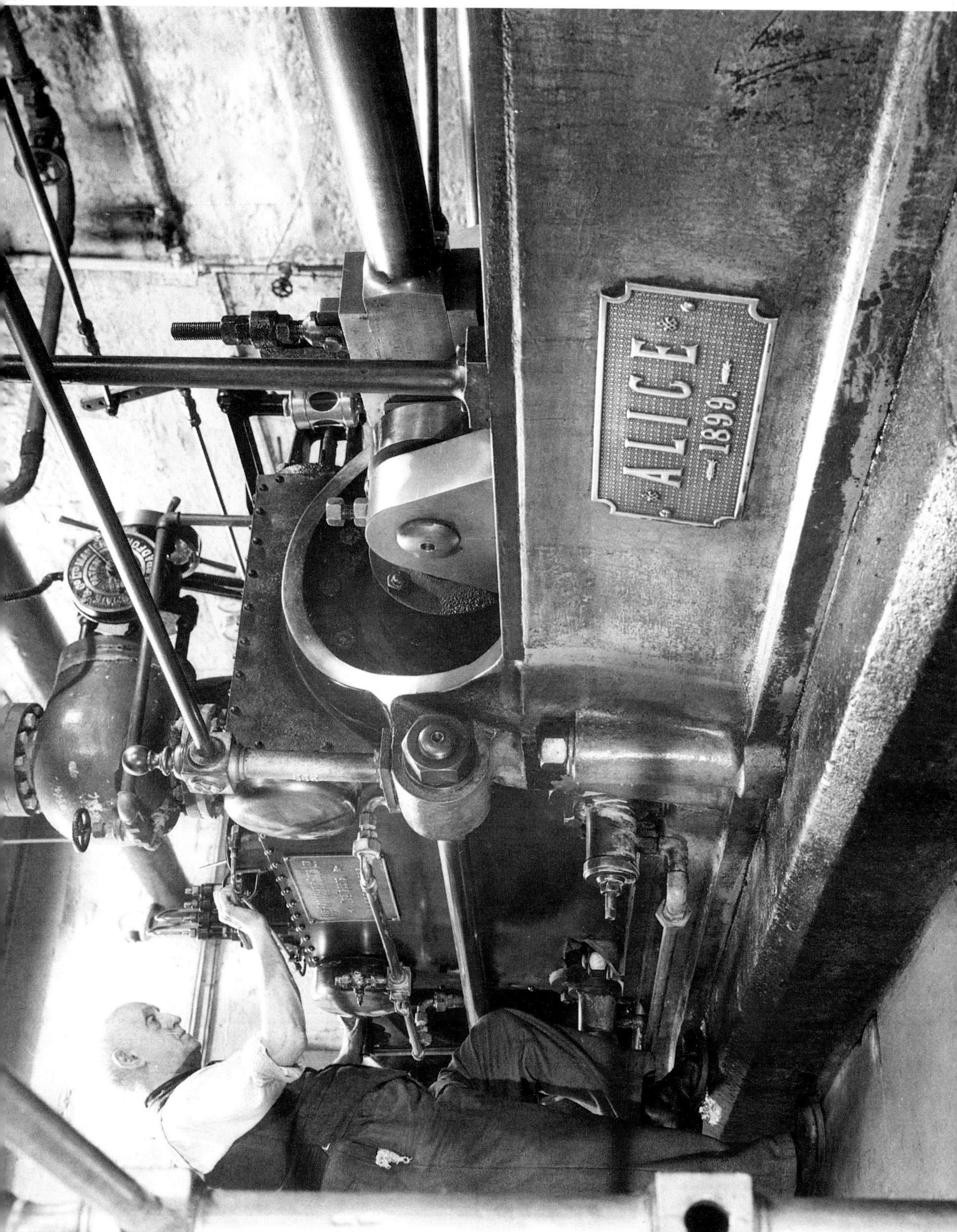
ALICE
1899

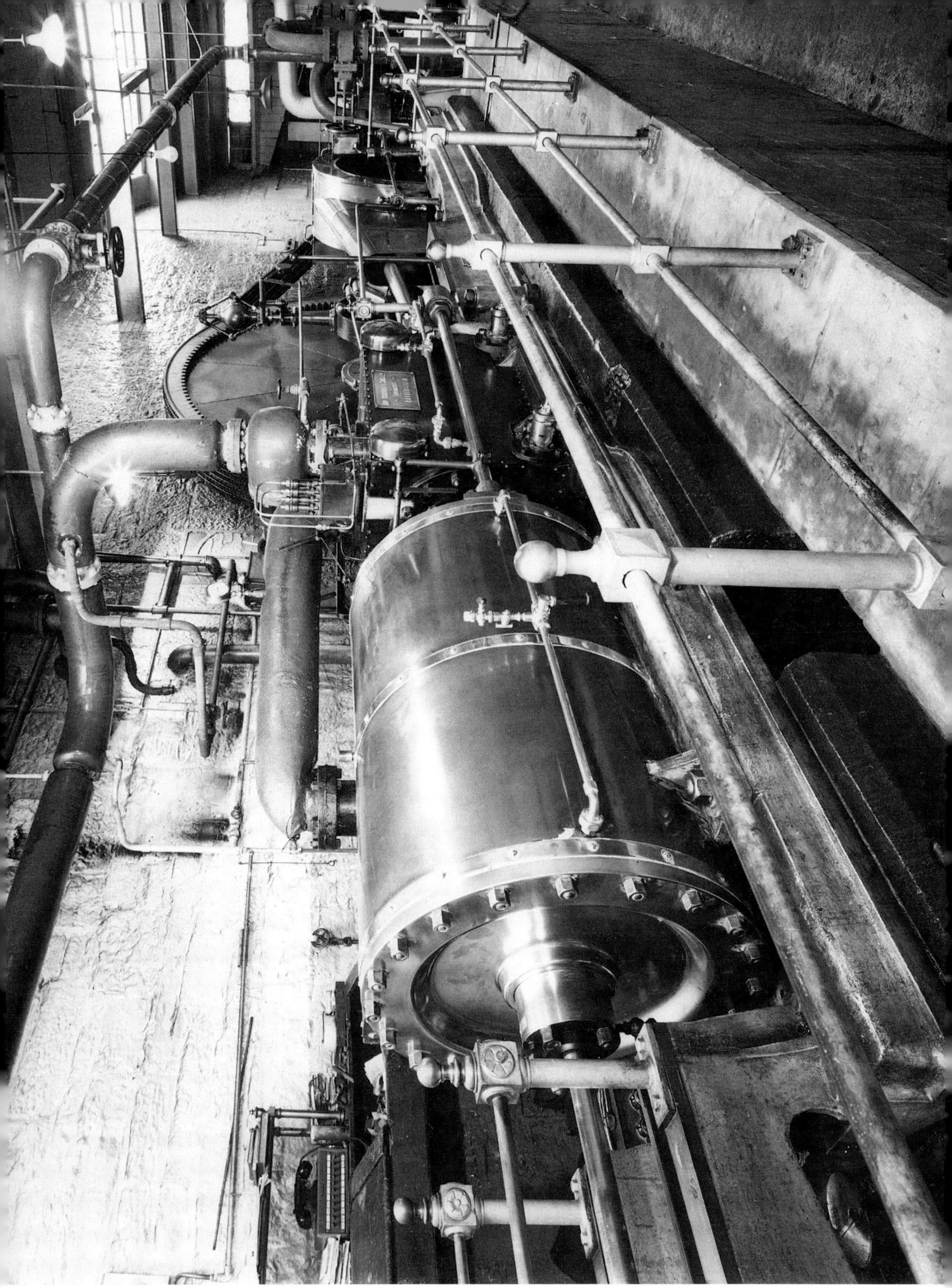

MAK/051 - Broad Weir Ironworks, Bristol.

1) Advertisements:
1893, for auction sale of the works. Page split on fold, needs conservation.

MAK/052 - Peter Brotherhood, Peterborough.

1) Advertisements:
1879, n.d. but c 1880, 1882 (2), 1887, 1888, 1892, 1894, 1895, 1896, 1897, 1903, 1903, 1905, 1910, 1911.

Reverse: 1897, C. A. Parsons, Heaton Works, Newcastle upon Tyne.

MAK/053 - Browett, Lindley & Co., Sandar Works, Patricroft, Manchester

1) Advertisements:
1887, 1891, 1893, 1894, 1903, 1900, 1901 (x2), 1905, 1910, 1911.

Reverse: 1900, Blechynden Boiler Co., Newcastle.
Reverse: 1901, Hudswell Clarke & Co., - Locomotives; Gutehoffnungshutte.

3) MS material:
2pp MS notes by GW about the firm.

5) Printed material:
Article from *Mechanical World*, 14 February 1901 about 'Sandon' compound engine made by Browett & Lindley.

Article from *Engineering* dated 25 March 1910 about 1500hp vertical high speed engine, 2pp of plates. Lacks text. (see p.49).

BROWETT, LINDLEY & CO., (1899) LIMITED.

PATRICROFT, MANCHESTER.

MANUFACTURERS OF

HIGH-SPEED SELF-LUBRICATING

STEAM ENGINES,

Two-Crank and Three-Crank Compound, and Three Crank Triple-Expansion Type,

— FROM —

40 TO 3000 HORSE-POWER.

ENGINES ON ORDER FOR THE FOLLOWING INSTALLATIONS:

CORPORATIONS OF—

Salford, Rochdale, Hull, Swindon, Kirkcaldy, Accrington, Middlesbro', Hornsey, Hendon, Beckenham, Colwyn Bay, Long Eaton.

ALSO FOR—

Trafford Park Power Company, Sir W. G. Armstrong, Whitworth and Co., Walter Dixon & Co., British Electric Traction Company, and many others, amounting to a total of over

35,000 HORSE-POWER.

LONDON OFFICE: 6, OLD QUEEN ST., WESTMINSTER, S.W.

Representative for Scotland—JOHN A. RUDD, 177, West George Street, Glasgow.

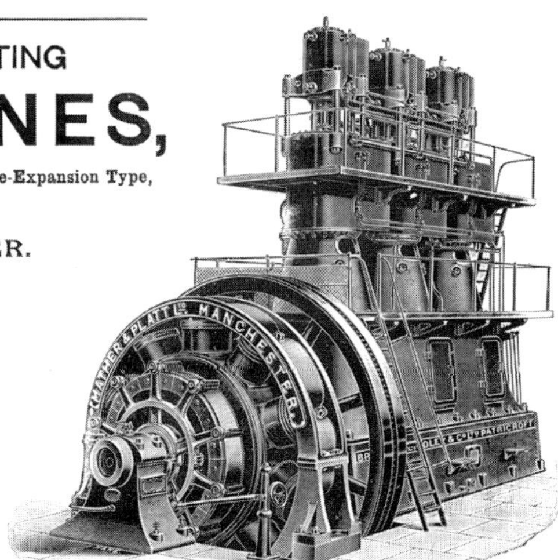

SALFORD CORPORATION—800 kw. PLANT.—100 R.P.M. 6377

1500-BRAKE-HORSE-POWER HIGH-SPEED ENGINE.

CONSTRUCTED BY MESSRS. BROWETT, LINDLEY, AND CO., LIMITED, MANCHESTER.

(*For Description, see Page* 391.)

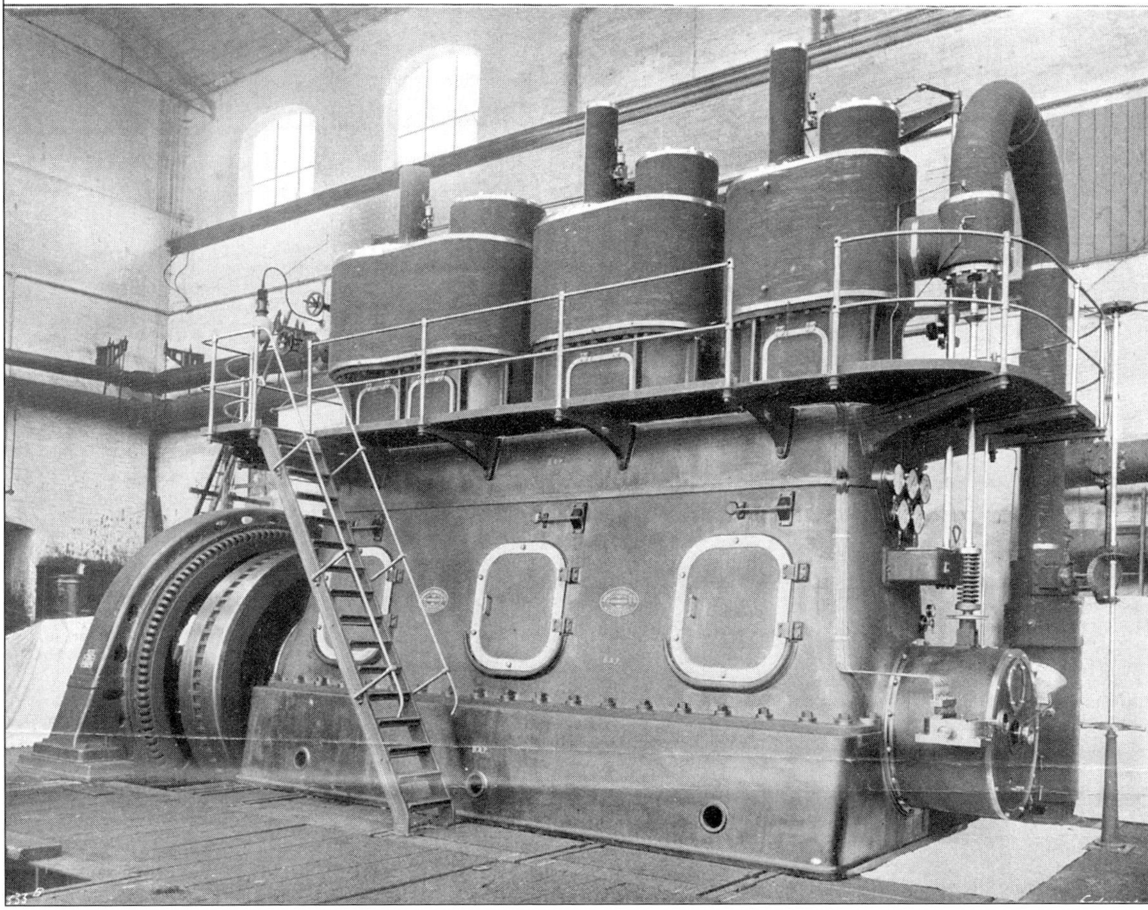

46 ENGINEERING. [DEC. 21, 1894.

BROWETT, LINDLEY & CO., LTD.,

PATRICROFT,

MANCHESTER.

MANUFACTURERS OF

HIGH-CLASS CORLISS ENGINES

SIMPLE, COMPOUND,

OR

TRIPLE EXPANSION.

ESTIMATES AND FULL PARTICULARS ON APPLICATION. 112

TELEGRAMS . . "SANDON, PATRICROFT."

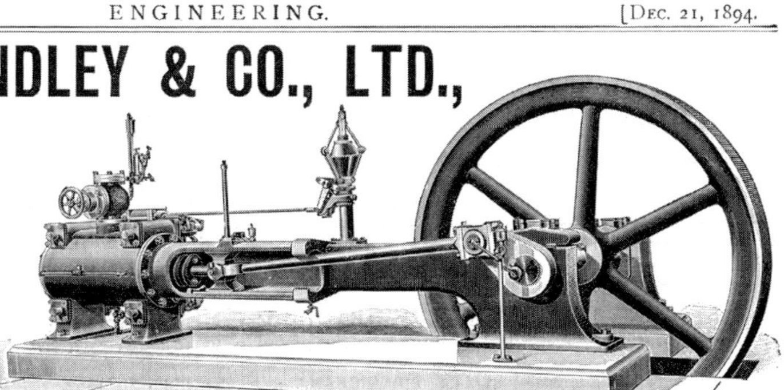

MAK/054 - Brown & May, Devizes, Wilts.

1) Advertisements:
1887.

Reverse: 1887 Saville Street Foundry and Engineering Works, Sheffield, - Hall's patent stone breaker and Baker's rotary pressure blower.

5) Printed material:
Photocopies of pages from *Steaming* illustrating various Brown & May traction engines. n.d. 5pp.

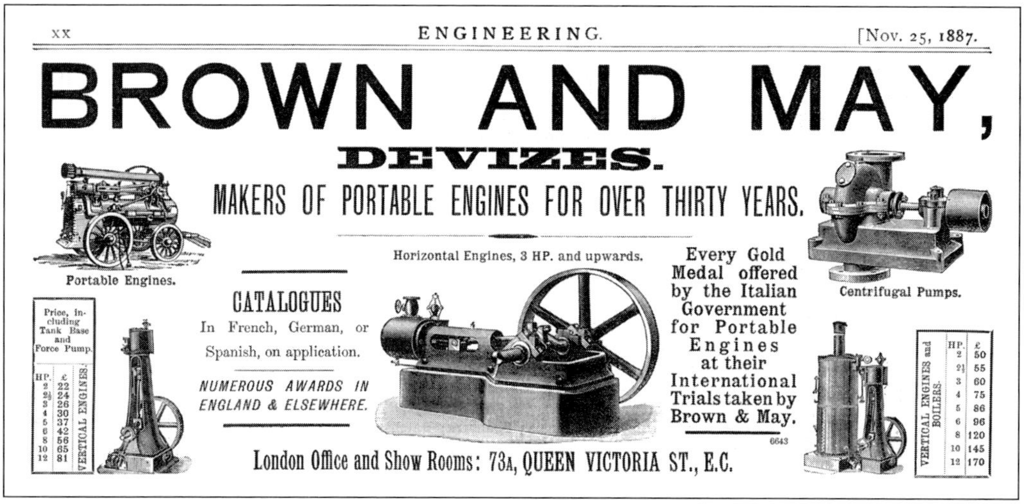

MAK/055 - Brush Electrical Engineering Co., Ltd, London and Loughborough.

4) Photographs:
8000 hp Brush engine and alternator at City of London Electric Lighting Co., 1903.
Photocopy of photograph of generator plant, headed 'Dover 1895'.

5) Printed material:
Article from *The Marine Engineer*, 1 March 1892, on ship lighting, with account of Brush plant.

MAK/056 - Buckley & Taylor, Castle Ironworks, Oldham, Lancs.

1) Advertisements:
1887, 1895,1896 (2), 1897 (2).

Reverse: 1887. Kendall and Gent, Manchester - machine tools .
Reverse: 1895 Crossley Bros, Openshaw, Manchester - oil engines
Same page: 1896 The Oldham Boiler Works Co., Ltd.
Reverse: 1897 Joseph Adamson, Hyde, Cheshire - boilers; The London Emery Works Co., - grinding machinery.

2) Letters:
MS letter to GW from James Wroe, 52 Sharples Hall Street, Oldham, Lancs, dated 6 November 1952 with information about firm's engines.

MS letter to GW from J. Yates, 2 Prospect Street, Hopwood, Heywood, Lancs, dated 27 July 1958. MS note by GW 'Hooley Bridge Towell Mills'.

MS letter to GW from Mrs Margaret Taylor, 80 Ronald Street, Clarksfield, Oldham, dated 10 October 1958. Refers to her late husband's collection of photographs, which he was welcome to inspect or she would send them on to him.

3) MS material:
Collection of MS notes by GW about the firm. 8pp.

4) Photographs:
The collection of photographs provides a valuable record of the firm's products, as well as useful information about workshop erecting practice. Most are unmounted. The catalogue is split into two parts: the first deals with identifiable engines or mills; the second with the rest. More investigation of the latter is required.

Identified engines or mills.
5 large prints showing replacement of part of the foundations of engine at Bluepitts Mill Castleton (James Orr & Sons Ltd) October 1918. Most with MS notes on back describing the print. Includes a 35 ton flywheel by RSJ's under rim and through spokes! (see p.55 [bottom] and p.56).

Lilac Mill, Shaw, about 1914. (see p.53).

Bangu (mill), 1890. This shows weaving sheds and a mill-chimney base in the course of construction. The background is very rural and looks vaguely oriental or South American.

2 prints of Lees and Wrigley No 5 mill being erected. 21 May and 4 June 1928. (see p.55)

Algurpe? Slate Corp. Horizontal compound mill engine 120 HP. n.d. , but c 1914.

Print of double helical gearing for Messrs J. Webb & Co., Ltd, Bury, March 1918.

Print of 41-in bore M P marine cylinder on large shaping machine. n.d. (see p.54b)

Print of broken valve box from intermediate cylinder, Oldham Albion Spinning Co., exploded 10 October 1930. (see p.54top)

3 prints of cylinder for P. A. Maccio, 92 contract, November 1925.

Print of 3-throw crank shaft for Don (mill) 1901.

2 prints of mill engine with MS note 'Croft or Clover' (see p.57-58).

2 mounted prints of engine at Unity Mill, Heywood.

2 mounted stereoscopic prints of inverted vertical marine-type pumping engines at Hyde Power Works, Pont Street, Manchester. 1 print of Buckley & Taylor engines, and the other of Mirrlees Watson, Co., Glasgow. n.d. , but c 1900. (see p.62-63).

Print of new high pressure Corliss cylinder for Pernambuco, 27 April 1910.

Science Museum print of plate from *Practical Engineer* 1 March 1895 of engine from Pine Mill. qv in 5) below.

Print of cast iron cylinder 9ft dia x 6ft deep. n.d.

2 prints each of 3-throw horizontal tandem pumps n.d. but 'War Work'. c 1914.

Envelope with 2 prints of engine at Pearl Mill, n.d. but c 1900. 1 repaired with sellotape. Wants conservation. With MS notes about the engine by GW., with Mr Walton, Engineer. (see p.60b & 61).

Print of New Victoria barring engine, July 1910.

Half a mounted print of vertical Corliss cylinder for Ferguson Bros, Carlisle, n.d. but c 1914.

Collection of 4 different prints of vertical triple engine for Regent or Globe Mill, n.d. but c 1914. (see p.59top).

Postcard with print of 2 Lancashire boilers made by Oldham Boiler Works for Oldham Gas Works, hauled by traction engine. Nice street scene, May 1909. (see p.60top).

Unidentified prints.
2 prints of part of foundry travelling crane, showing boiler and part of engine.

Print of ship's winch, dated May 1918.

Postcard of centrifugal pump, n.d.

Print of fractured shaft.

2 prints of unidentified machines with cast name 'Kynoch, Birmingham'. One is dated 16 May 1916. They both appear to be driven by flat belting and may possibly be food processing plant.

Print of small horizontal Corliss mill engine, with pair of vertical pumps mounted at side.

Print of very large gear ring being turned on dogged faceplate, n.d. but c 1914.

4 prints of valve gear parts, and cylinder, n.d.

Print of gearing, retouched for reproduction, n.d.

4 copies of print of Corliss engine being erected.

2 prints of large Corliss cylinder.

2 copies of postcard of Corliss valve gear.

Print of unidentified casting on horizontal boring mill, n.d.

3 prints of large Corliss mill engine part erected, n.d.

Dyeline photoprint of Corliss valve gear, n.d.

5) Printed material:
Leaf from unknown periodical, 1893 with brief obituary of James Taylor, partner in the firm and who died 10 November 1892.

Patent specification (90 of 1879) for Buckley & Taylor's toothed wheels, 4pp illus.

Article from *Engineering* 26 August 1887 about horizontal compound tandem mill engine at Gordon Mill, Oldham.

Photocopy of page from *The Practical Engineer*, dated 1 September 1893, with indicator diagrams from Leesbrook Spinning Mills, Lees. Presumably by B. & T.

Article from *The Practical Engineer*, 5 May 1893 with account of christening ceremony of the engines of Pearl Mill, Oldham.

3 Photocopies of article from *The Practical Engineer,* 1 March 1895, with account of 1800 HP compound mill engine at Pine Mill, Oldham.

Article from *Mechanical World,* dated 23 October 1896 describing 1600hp compound beam engine at mill at Halifax owned by Rippondale Commercial Company Ltd. With glossy photographic copy of above.

Article from *Mechanical World*, dated 31 December 1897, about 1500hp compound beam engine for Hathershaw Spinning Co.,

Newspaper cutting from *Oldham Chronicle*, dated 17 April 1971 about 2 hours' notice of redundancies of 31 Buckley & Taylor staff.

6) Trade literature:
Folder with collection of leaflets, no date but c 1915. 11 pp.
Duplicate leaves from above. 5pp

7) Other material:
Collection of photoprints of drawings of Nile Mill, Oldham designed by John Tattersall. Dated August 1898. Much reduced from originals. 9 sheets. With collection of 4 photocopies of side elevation.

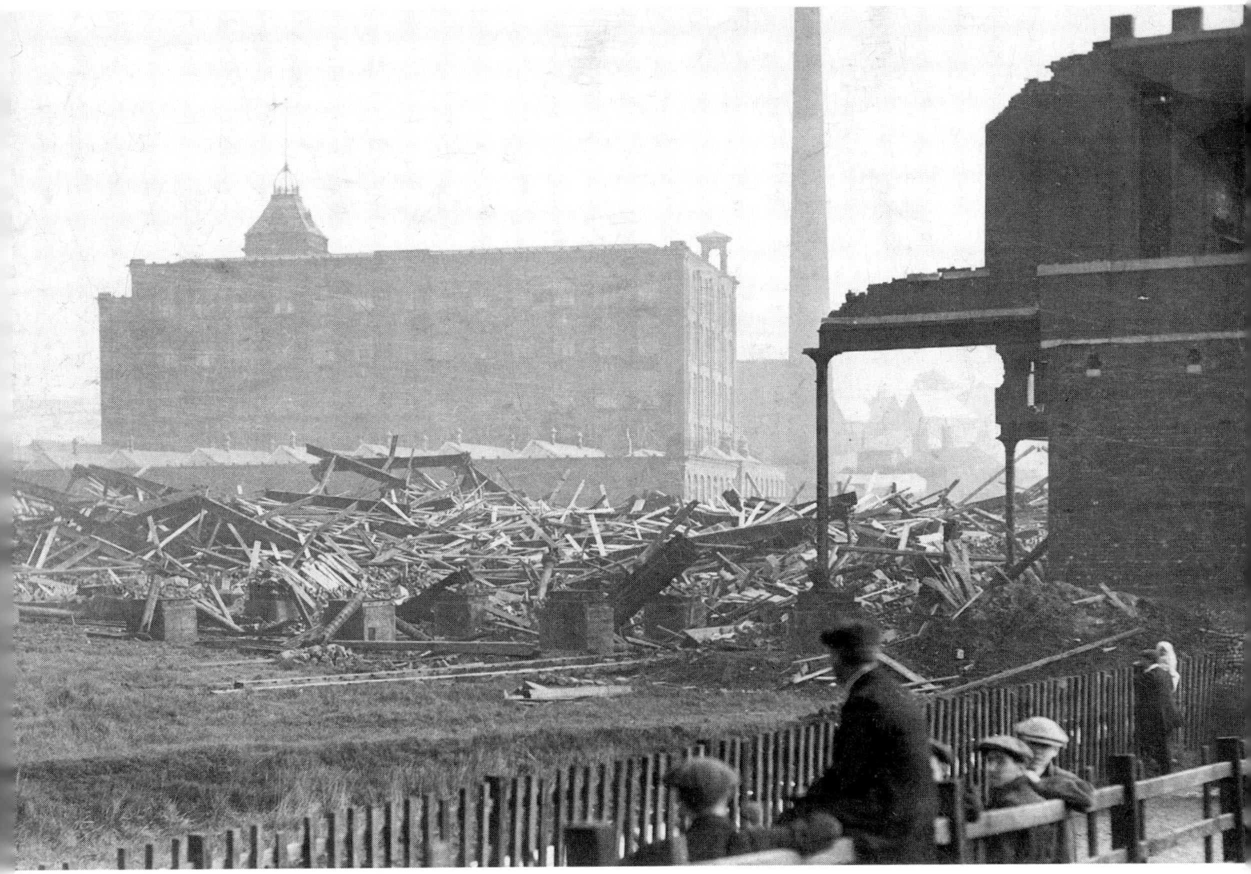

Lilac Mill, Shaw, c 1914

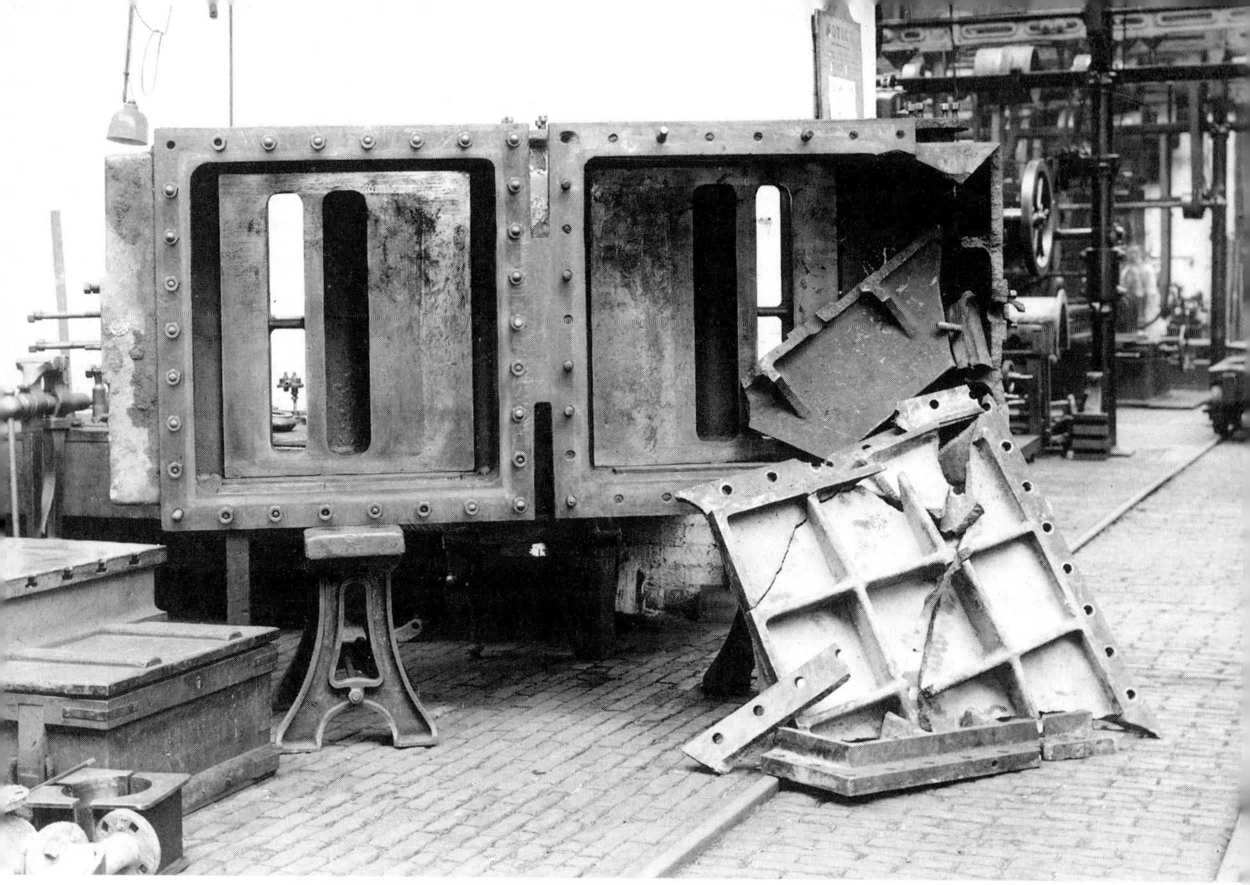

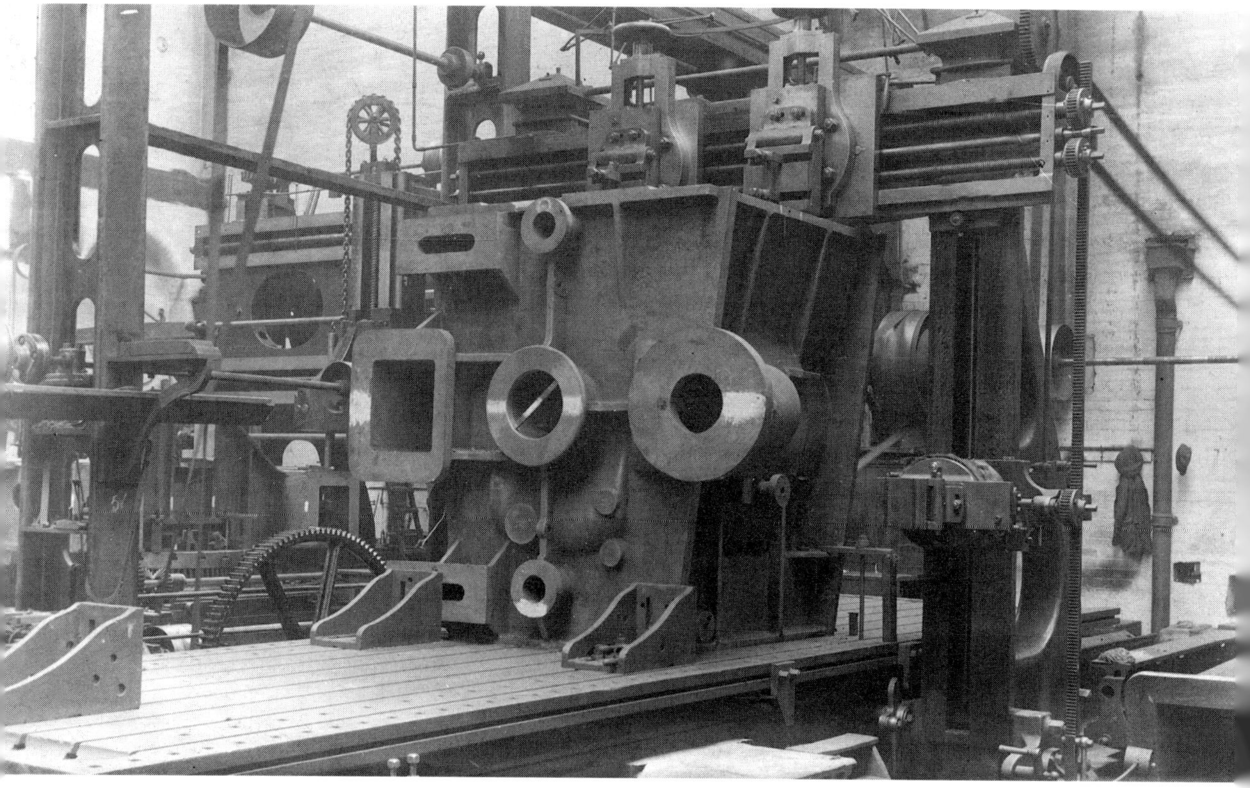

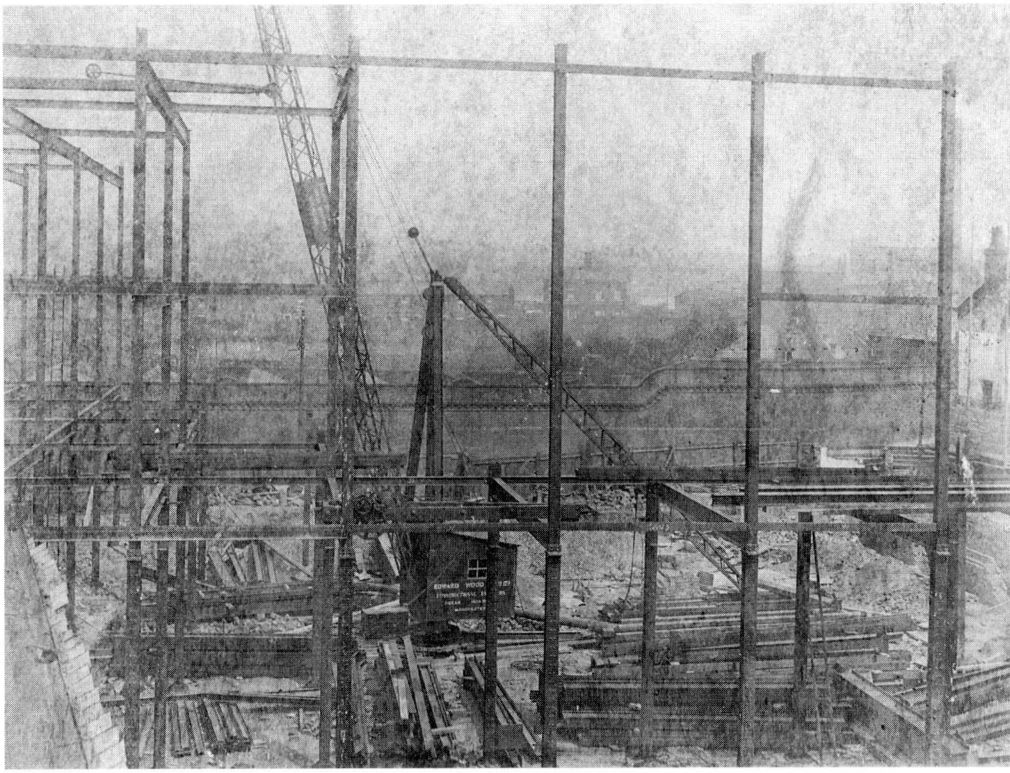

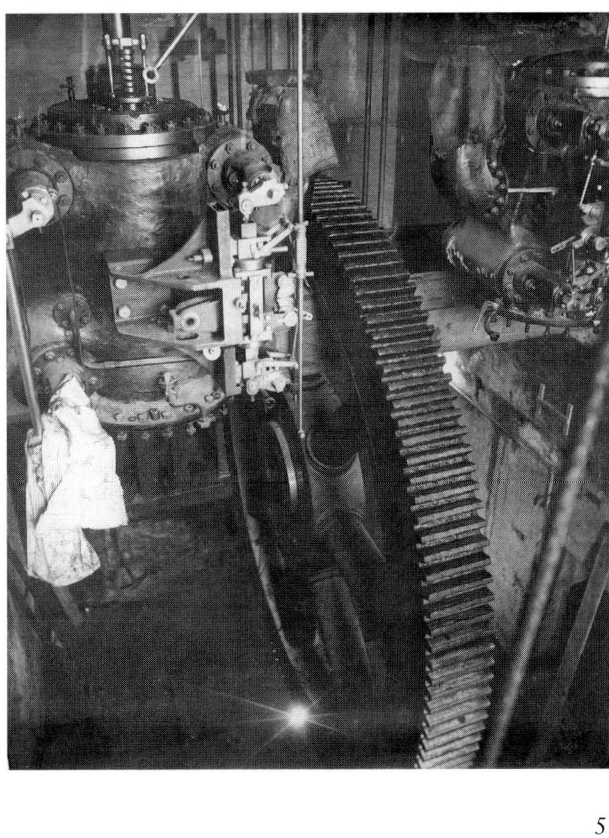

Opposite page top: Oldham Albion Spinning Co., 1930 (valve box after explosion).

Opposite page bottom: 41-in bore mp marine cylinder on large shaping machine (n.d.).

Above: Lees & Wrigley No5 mill being erected, 1928.

Right: Bluepitts Mill, Castleton, (J. Orr & Sons, Ltd) showing replacement of part of foundation for engine, 1918.

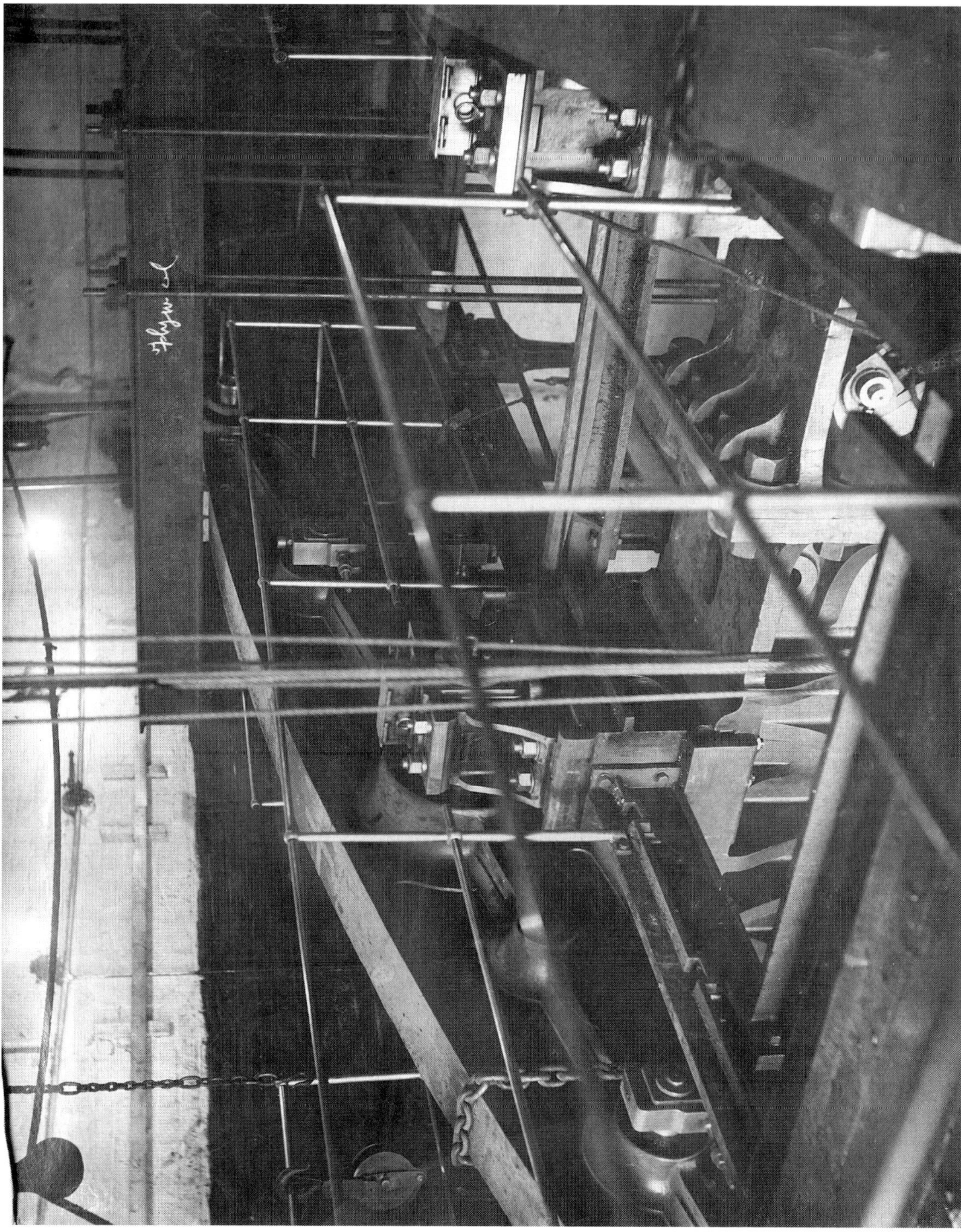

Above: Bluepitts Mill, Castleton.

Opposite page & p. 58: The Buckley & Taylor engine made for Clover Mill, Rochdale in 1906.

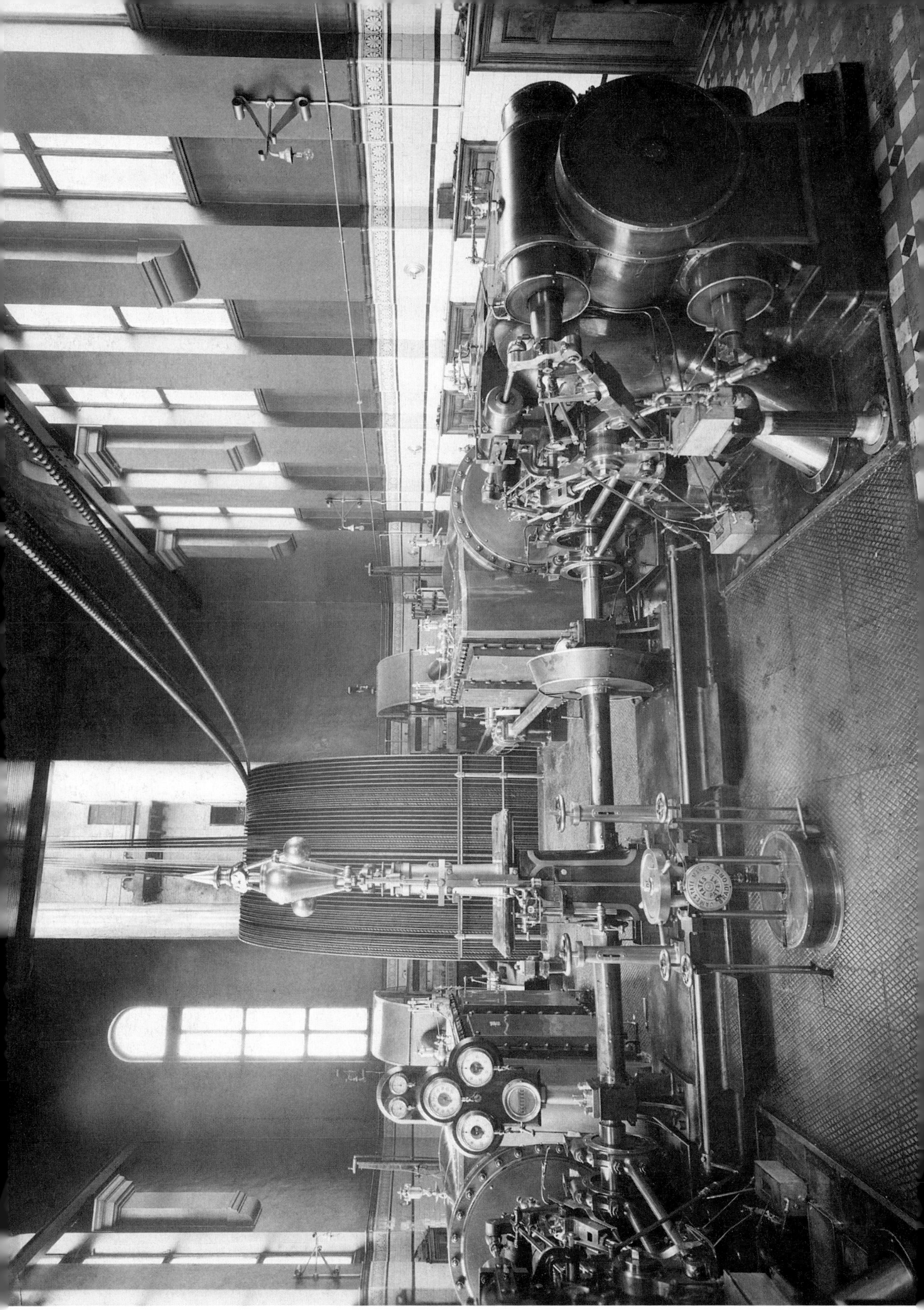

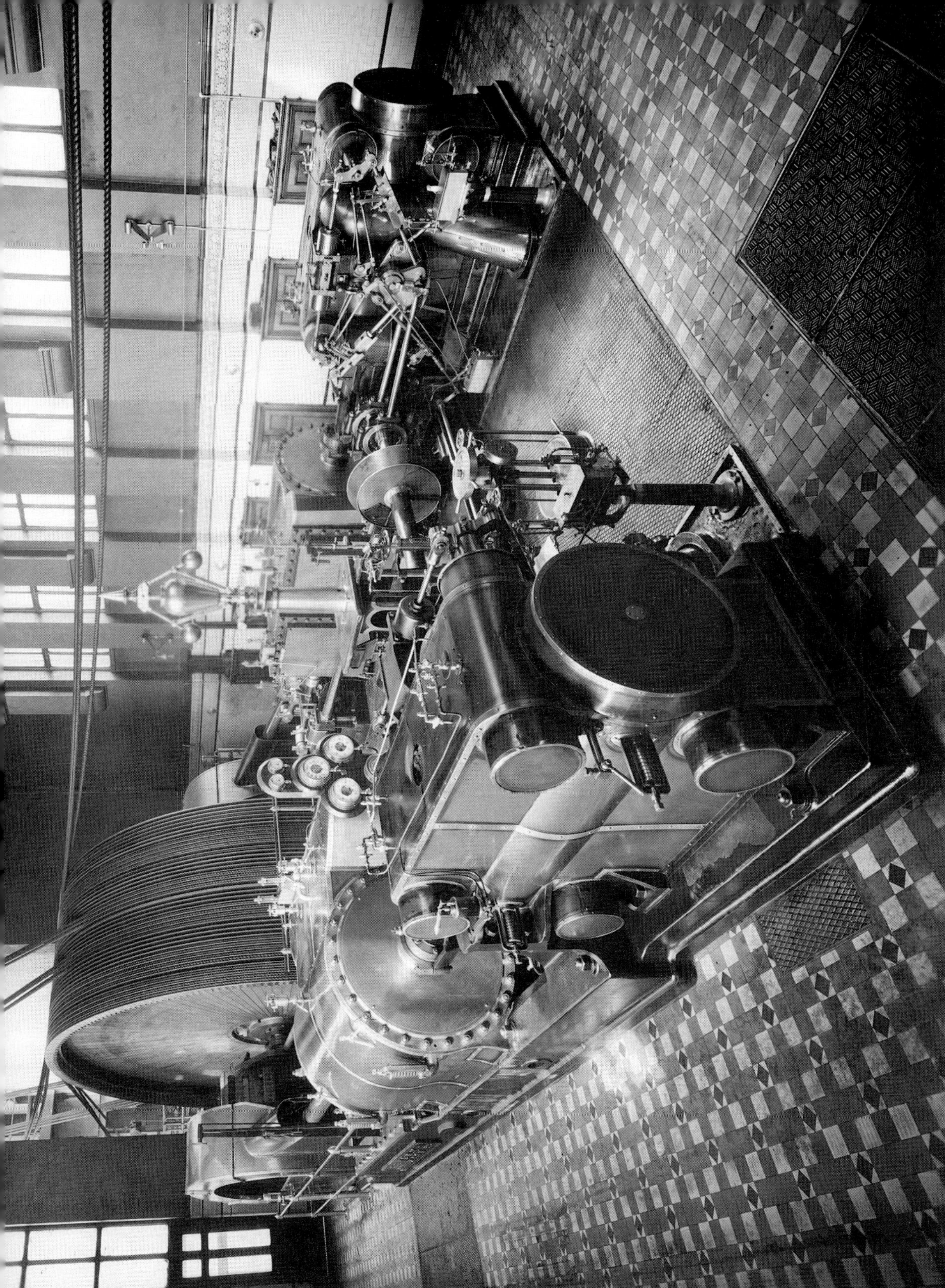

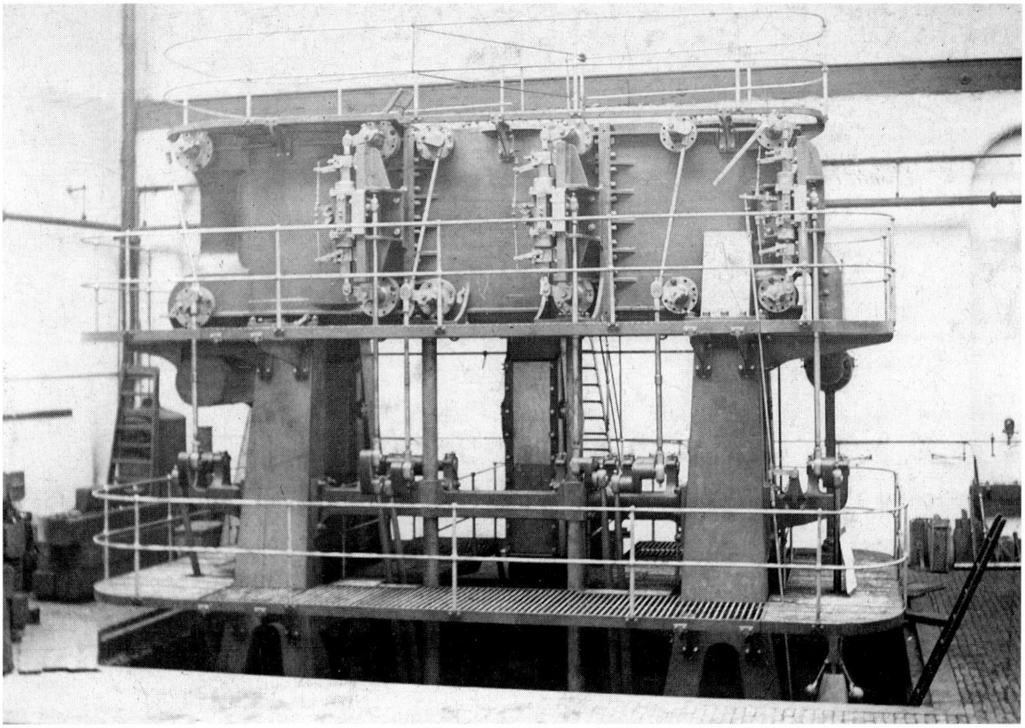

Above: Vertical engine, Regent or Globe Mill, c 1914.

Below: Celgwyn Slate Coy. hp-12in; lp-22in. Stroke 2ft-3in; Revs 100; Boiler pressure 100.

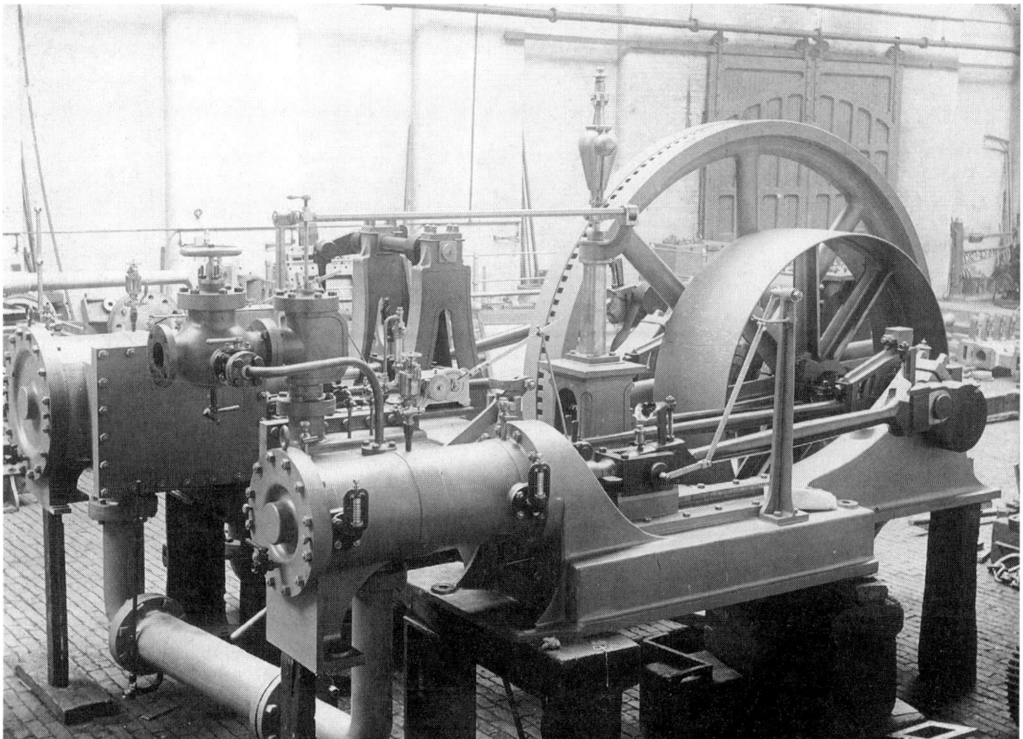

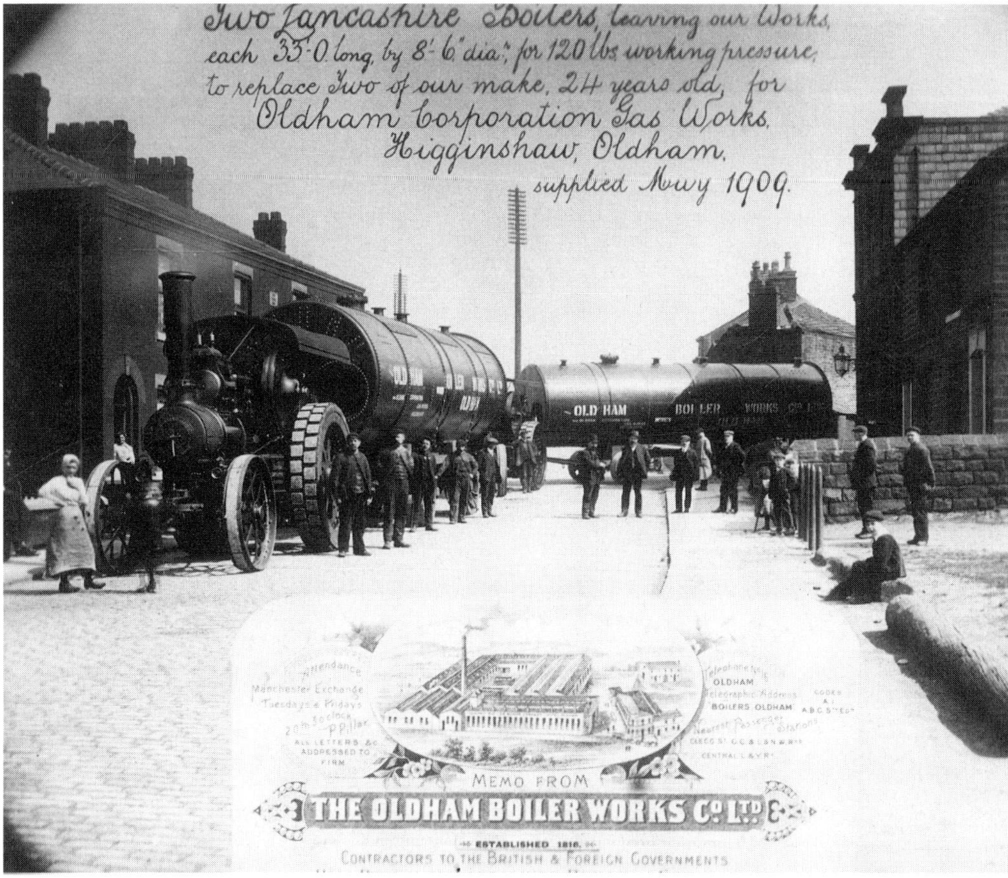

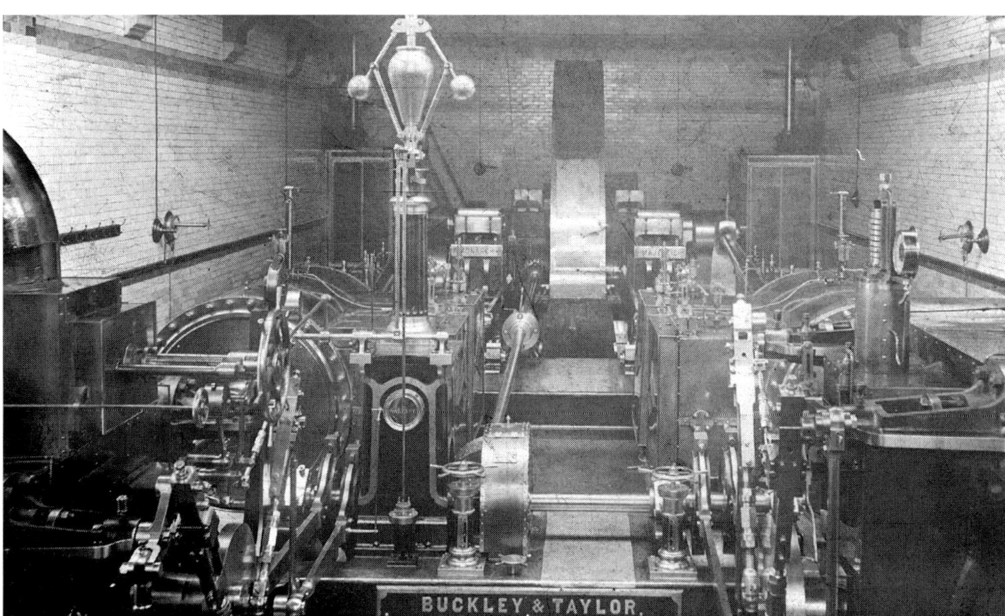

Pearl Mill, Oldham.

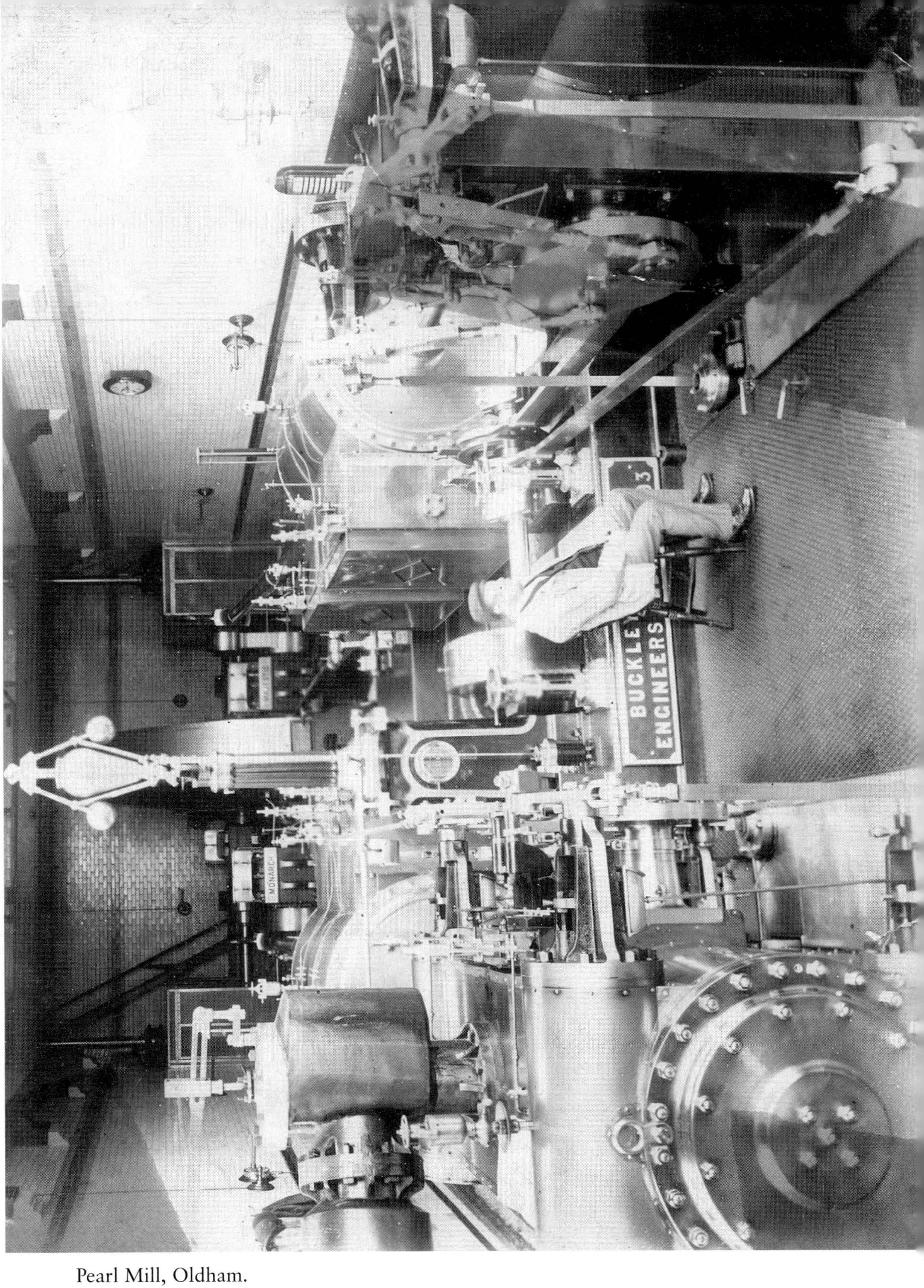

Pearl Mill, Oldham.

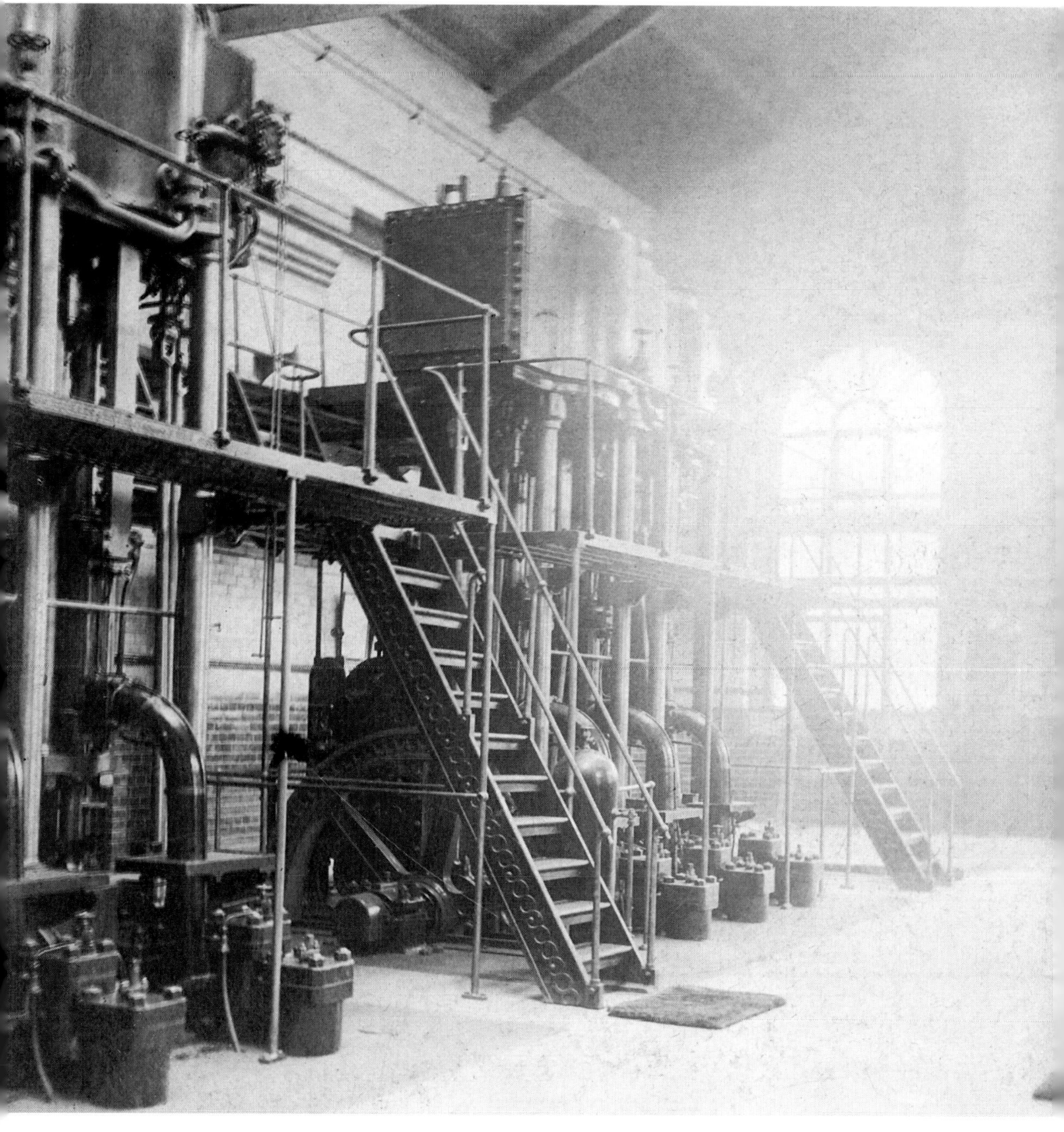

Hyde Power Works, Manchester.

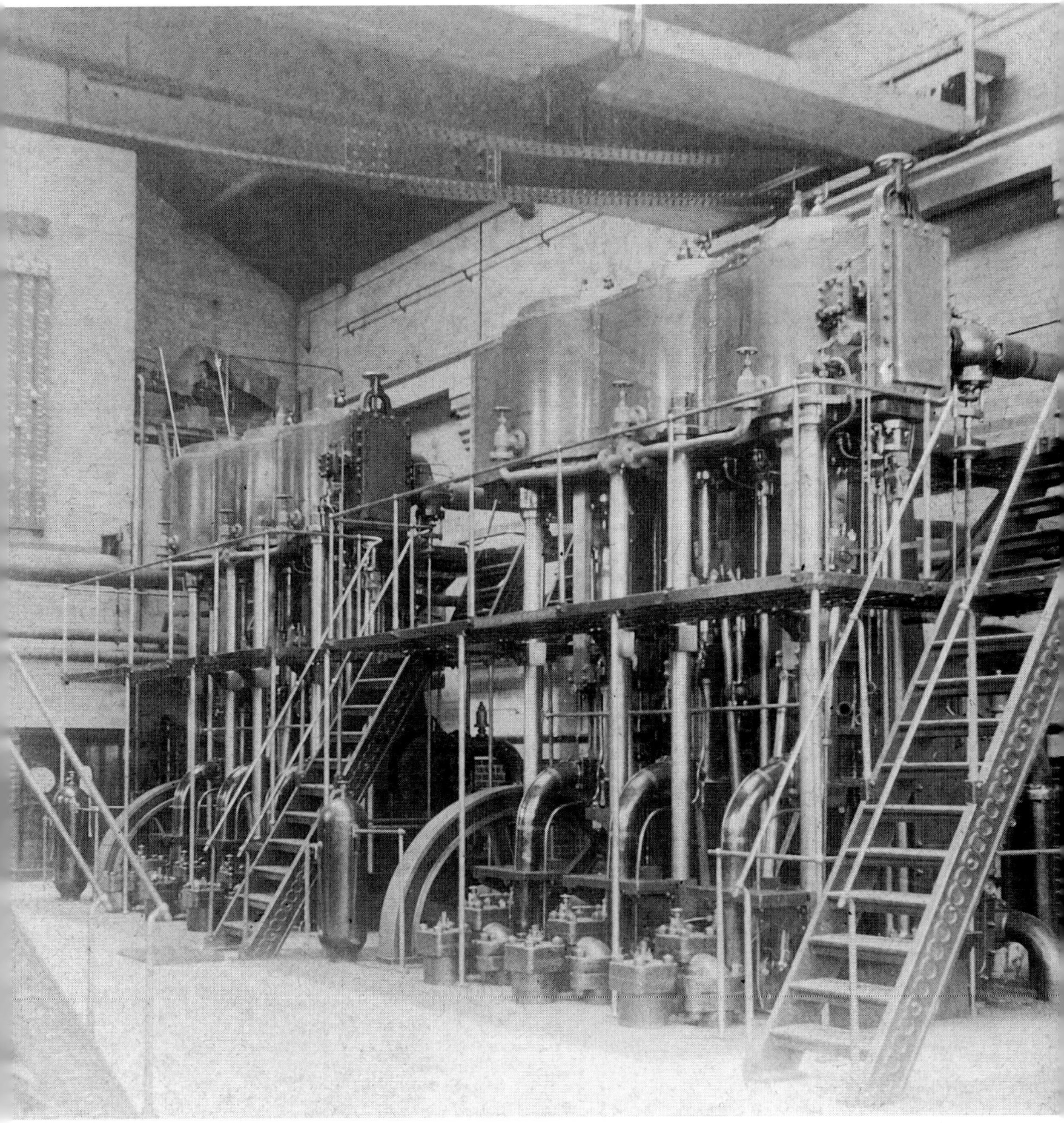

Hyde Power Works, Manchester.

MAK/057 - Bumstead & Chandler, Cannock Chase Engine Works, Hednesford, Stafford.

1) Advertisements:
1869, 1886 (2), 1893, 1894 (2).

3) MS material:
MS notes about the firm by GW 1 p.

5) Printed material:
Pages from Header & Powles, *A Handbook on the Steam Engine,* 1896 about their engines, pp 350-353.

Pages from Norris & Morgan, *High Speed Steam Engines,* pp 90-93, with B&C advertisement on p xii.

MAK/058 - Burnley Ironworks, Burnley.

2) Letters:
Letter to GW dated 14 July 1965 from Burnley Paper Works Co., about the Burnley Iron Works engine in their works.

3) MS material:
MS notes about Burnley Iron Works by G W.
MS notes about Marslands, with photocopy copy ?ex Dr Collier.

4) Photographs:
Print of horizontal Corliss tandem mill engine at Slater's, Calderdale Mill, Burnley. (see p.65).

5) Printed material:
Photocopy of page from *Model Engineer* 12 April 1951 with note about original drawings of engine at Tregollon mine, Cornwall, 1863, by Burnley Iron Works.

Article from *Mechanical World* 23 May 1891, about horizontal mill engine at Messrs O. & J. Folds, Trafalgar Mills, Burnley.

Photocopy of page from Nasmyth, *Cotton Mill Construction*, 3ed, illustrating twin horizontal mill engine.

Opposite page: Burnley Ironworks tandem engine, Calderdale Mill, Burnley

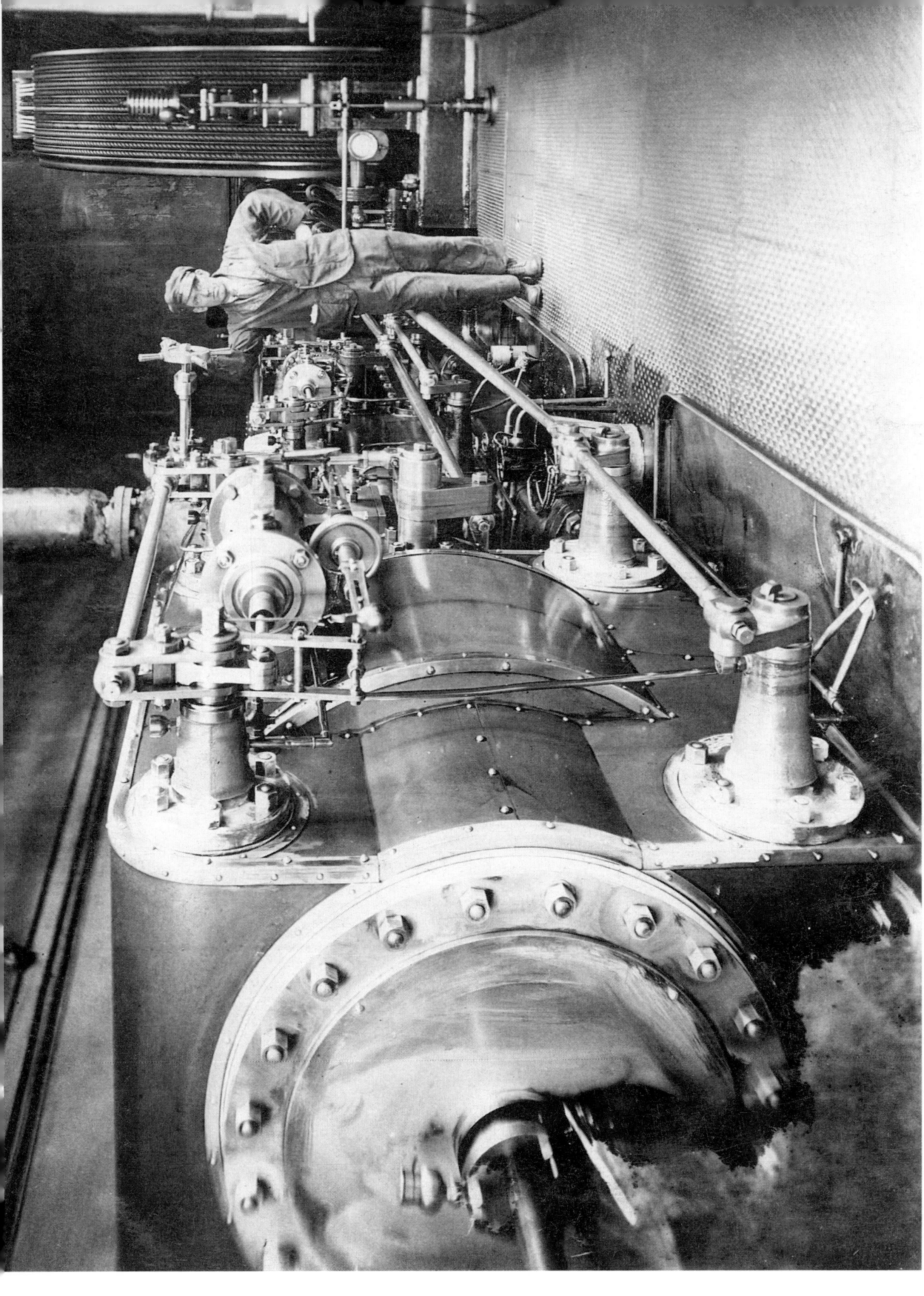

MAK/059 - Butters Brothers, Percy Crane & Engine Works, Lamphill Street, Paisley Road, Glasgow.

1) Advertisements:
1882.

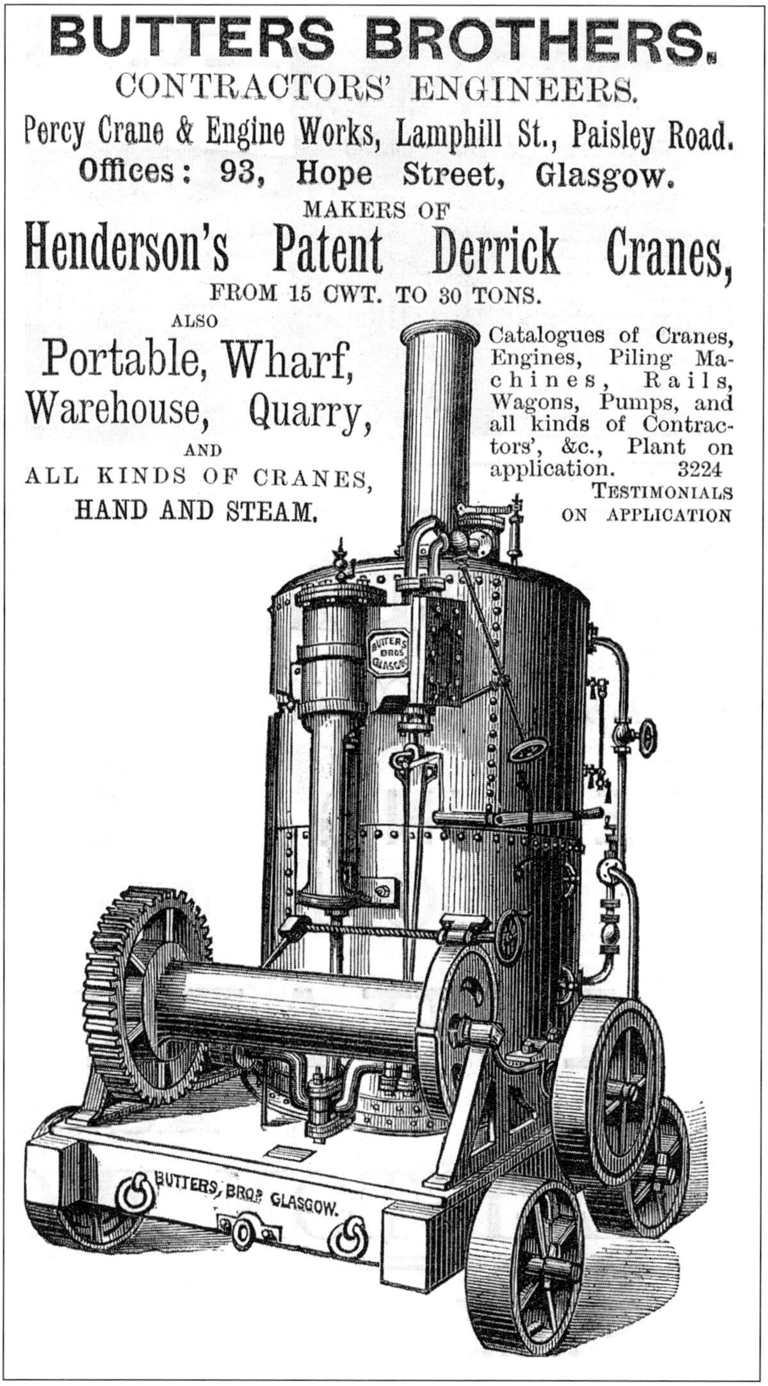

MAK/060 - Cail, Denain Nord (Société Française des constructions méchaniques).

4) Photographs:
Print of tandem piston valve horizontal engine, neg. 11768. (see below).
Print of Corliss horizontal engine, neg. 11791.

7) Other material:
Scale drawing of tandem horizontal above dated 1951.
Scale drawing of Corliss above dated 1950.

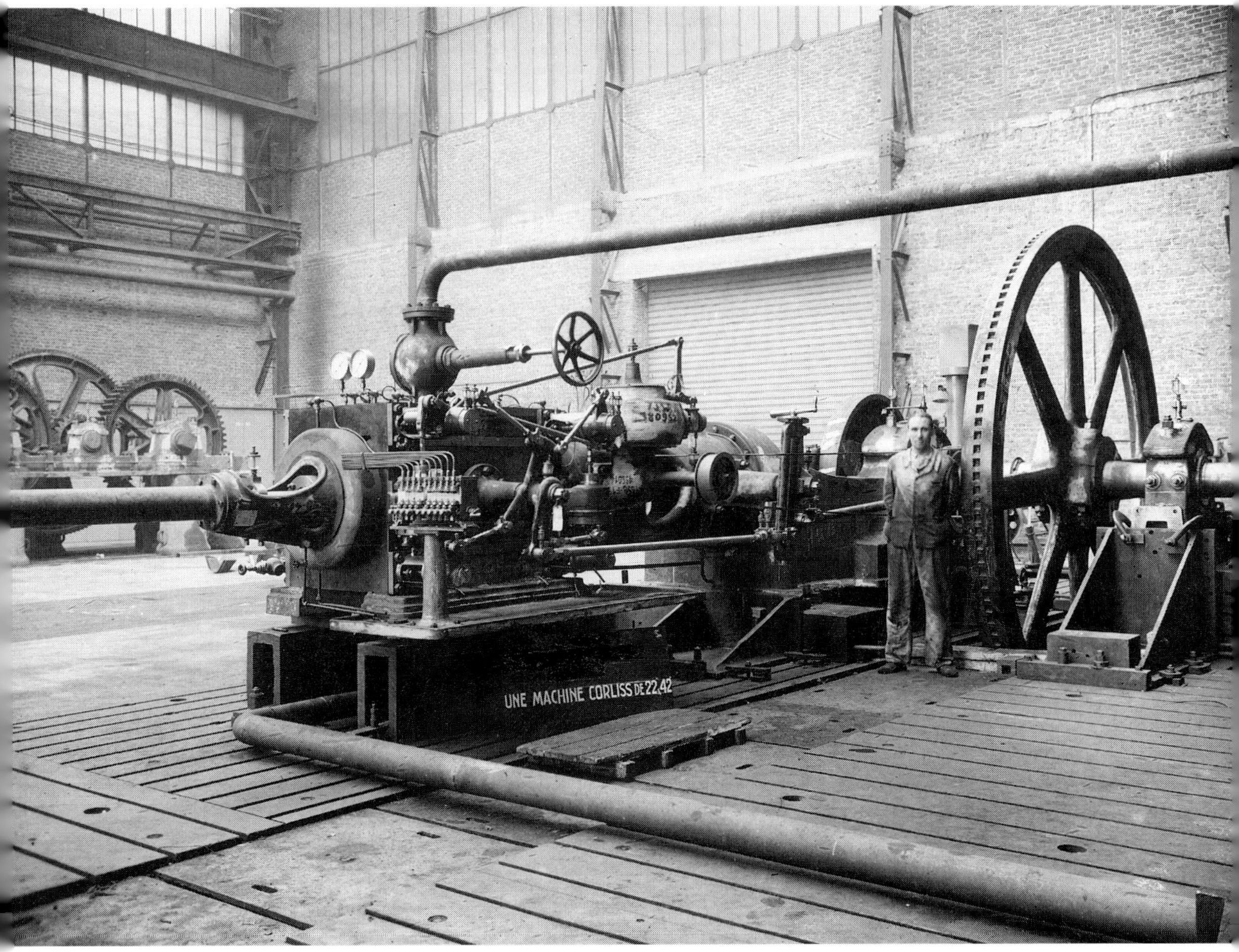

MAK/061 - Calvert & Co., Hope Foundry, Folly Hall, Huddersfield.

5) Printed material:
Photocopy of page from *Industries of Yorkshire* part 2, 1892, p 82, with account of the firm's activities.

MAK/062 - John Cameron, Oldfield Road Ironworks, Salford, Lancs.

1) Advertisements:

1887 (see below), 1896 (2).

Reverse: 1896 Thomas Piggot & Co., Atlas Works, Birmingham. - welded pipes.

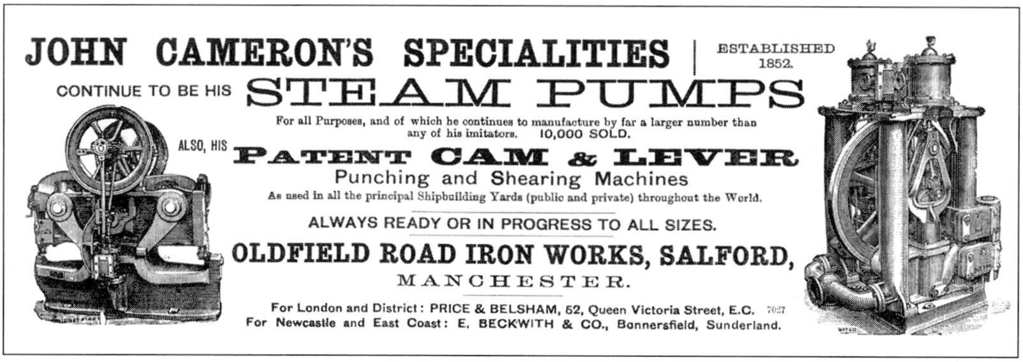

MAK/063 - Campbell & Calderwood, Soho Engine Works, Paisley, Scotland.

1) Advertisements:

1894 (see below), 1897, 1905 (2), 1910, 1911, 1913.

Reverse: 1910. James Simpson, Newark-on-Trent.

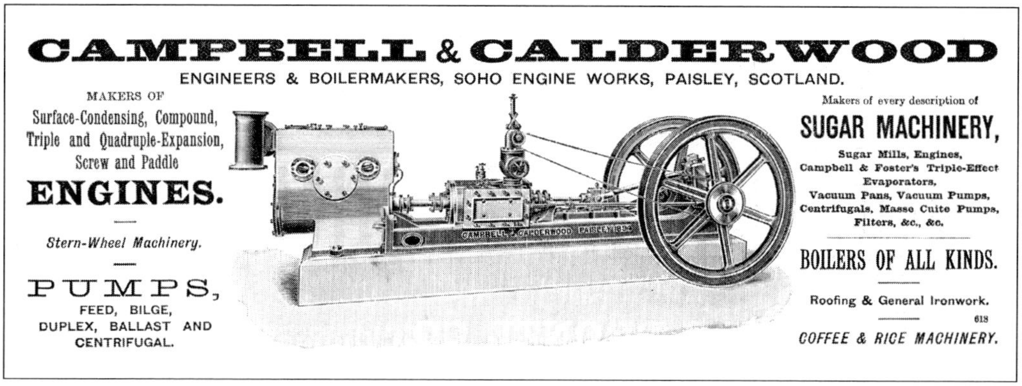

CAMPBELL & CALDERWOOD,

ENGINEERS AND BOILERMAKERS,

SOHO ENGINE WORKS, PAISLEY, SCOTLAND.

MAKERS OF

SURFACE-CONDENSING, COMPOUND, TRIPLE AND QUADRUPLE-EXPANSION, SCREW AND PADDLE **ENGINES**

STERN-WHEEL MACHINERY.

PUMPS FEED, BILGE, DUPLEX, BALLAST AND CENTRIFUGAL.

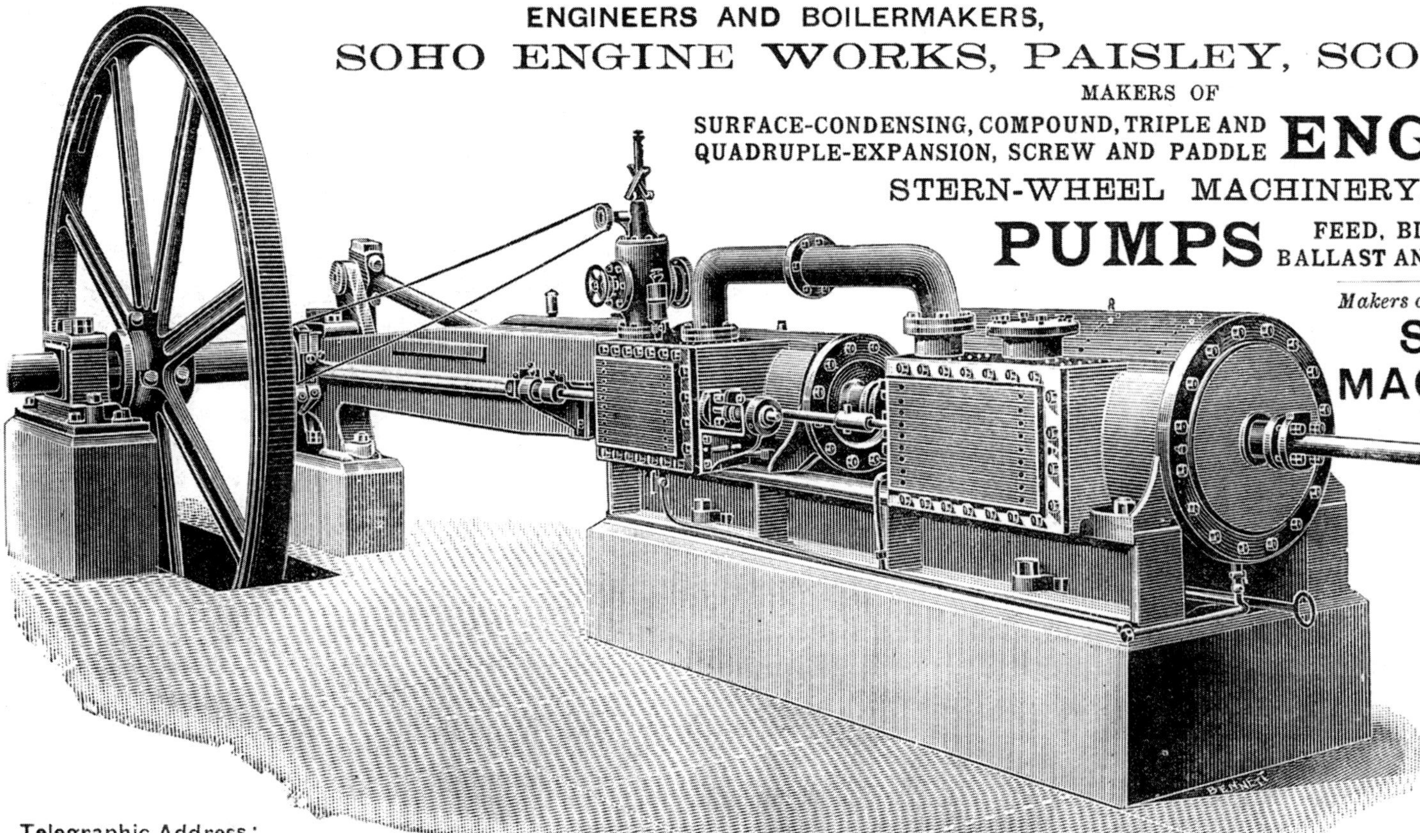

Makers of every description of

SUGAR MACHINERY

Sugar Mills,
Engines,
Campbell & Foster's
Triple - Effect
Evaporators.
Vacuum Pans,
Vacuum Pumps,
Centrifugals,
Masse Cuite Pumps
Filters, &c., &c.

BOILERS

OF ALL KINDS.

Roofing and
General Ironwork

Telegraphic Address:
"SOHO, PAISLEY."

LONDON AGENT: J. W. CLARIDGE, 39, Victoria Street, S.W.

2268

MAK/064 - Carels, Ghent, Belgium.

1) Advertisements:
1921.

Reverse: 1921 Cochran & Co., Annan - vertical boiler.

3) MS material:
MS notes by GW, 1 p.

5) Printed material:
Article from *Engineering* 20 July 1900 with account of 1000 hp horizontal engine and generator at Paris Exhibition.

2 copies of booklet: *A visit to Carels' works, Ghent* A reprint of articles from *The Engineer* 5 and 19 May 1922 20pp [1 copy lacks wrapper].

MAK/065 - James Carmichael & Co., Ward Foundry, Sessions Street, Dundee.

2) Letters:
Letter to GW from David Elder, 183 Overgate, Dundee, dated 16 November 1953. 2pp.

Letter to GW from H&A Scott Ltd, Tayfield Works, Dundee, dated 5 January 1954 about their Carmichael engine. l pp.

Letter to GW from John Carmichael, Boghaugh, Dalginross, Comrie, Perthshire dated January 1954 1 pp.

Letter to GW from Dundee Museums and Fine Art Galleries, dated 8 January 1954, 3pp.

Letter to GW from H. Mitchell, 5 Fingask Street, Dundee, dated 17 January 1954. Mentions having a list of Carmichael's engines. Mentions hearing that Mr Gorric, (sic) Carmichael's photographer, was arranging to send GW a selection of photographs. (See 4 below).

Letter to GW from C. W. Proctor, Naffiali Jute Mill, Hazinagar PO, West Bengal, dated 14 February 1954. Says his father had been Carmichael's company secretary and wrote booklet: *The House of Carmichael* [See 6) below]. Says his mill had a 180hp Carmichael engine, built in 1905.

Letter from Dundee Public Libraries, dated 27 February 1954, thanking him for photographs. Also mentions Mr Corrie, (sic) see 5 above.

3) MS material:
MS note by GW about John Carmichael in above.

4) Photographs:
This is a fine collection of prints illustrating mostly manufacturing and erecting techniques. Most are undated, but all are c. 1920. Nos 1-13 are in contract order.

Print of flywheel end of horizontal mill engine, contract no 34, MS note 'in Calcutta'

2 prints of uniflow engine, 18in stroke, Contract no 180. (see p.74).

Print of uniflow engine, 18in stroke, Contract no 181.

Print of sectioned drawing of uniflow engine, 15in dia x 18in stroke, Contract no 238.

 Opposite page: Corliss engine, J. Carmichael & Co.

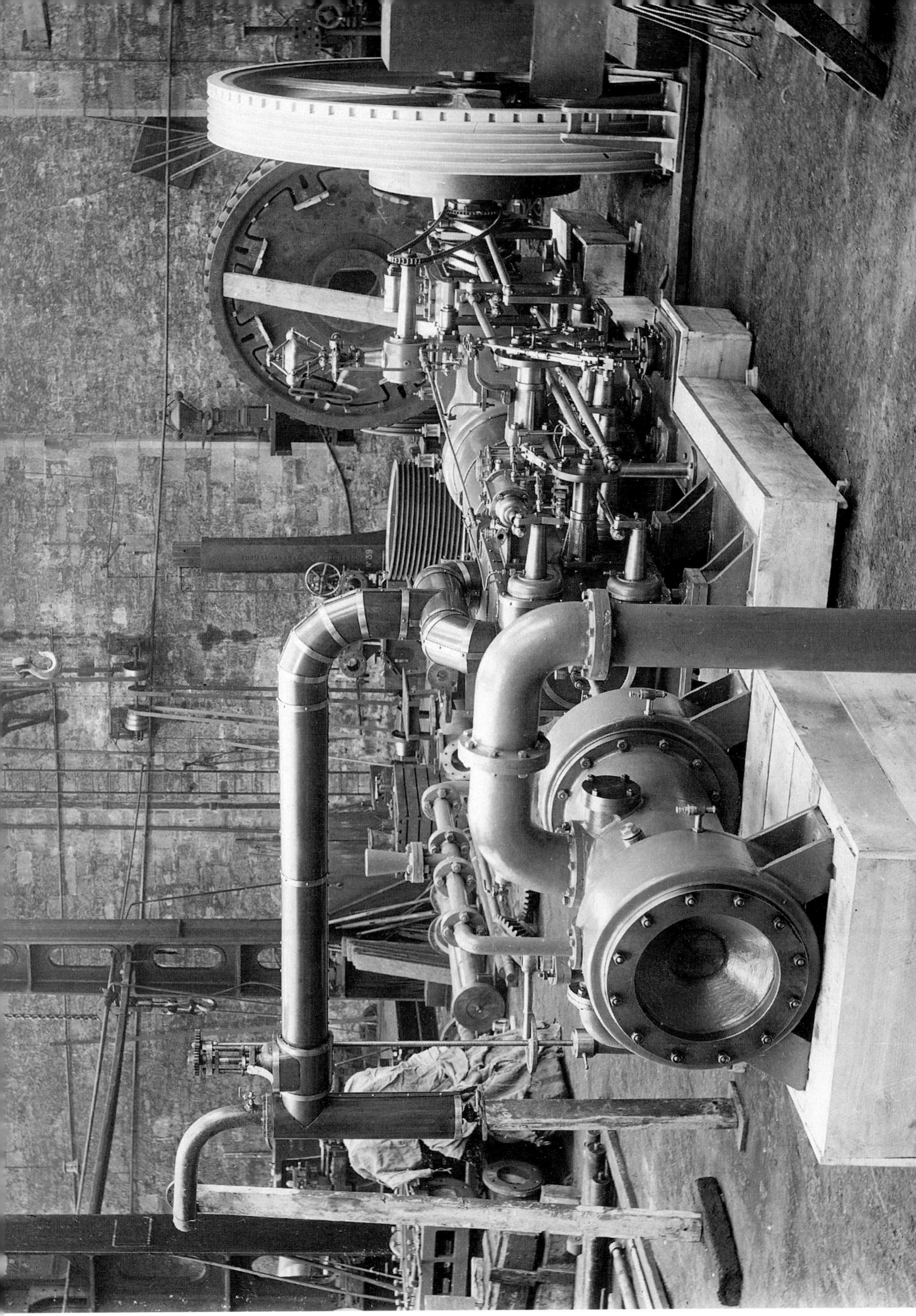

MS note 'Ca 1920'.

Print of cutting the keyway in a segment of a 80 ton rope flywheel. Contract no 253. Uses a vertical slide on large shaping machine.

Print of fitting the arm of a segment of a 25ft dia. rope flywheel. Contract no 254.

2 prints of building a 90 ton rope flywheel in pit, for turning. Contract no 257b.

Rope flywheels, one comprising 10 segments, carrying 44 2in ropes, weight 90 tons, diameter 25ft. One with 8 segments, 14ft diameter, 8 $1^5/_8$ in ropes, weight 12 tons. Contract no 258a.

Print of rope flywheel and crankshaft for 2500hp engine, 25ft dia weighing 120 tons, grooved for 22 2in ropes. Contract no 258b.

Print of shaft and webs weighing 22 tons for 2500 hp engine, Contract no 258c.

Print of low pressure cylinder, Contract no 270.

Uniflow steam engine with drop valve gear. Contract no 315.

Single reduction steam turbine. Contract no 323.

Other prints
Print of flywheel end of large horizontal mill engine. MS note 'In Calcutta' (see p.74).

Print of model side-lever marine engine [Ex Dundee museum. See letter of 8 January above. APW]

Print of governor and hp valve gear of 1,000hp cross compound Corliss engine at Calcutta.

Print of valve bonnet and details for uniflow engine. Contract no. not recorded.

Print of turning 20ft balance flywheel. Contract no. not recorded.

2 prints of horizontal Corliss engine with MS note Winter's Boat Factory, Carnoustie (see p.73).

Print of large gear and shaft.

6) Trade literature:
Booklet: *The House of Carmichael*, n.d. , 32pp. Note: some of the prints in 4) above appear in this booklet.

Opposite page: Carmichael engine in Calcutta.

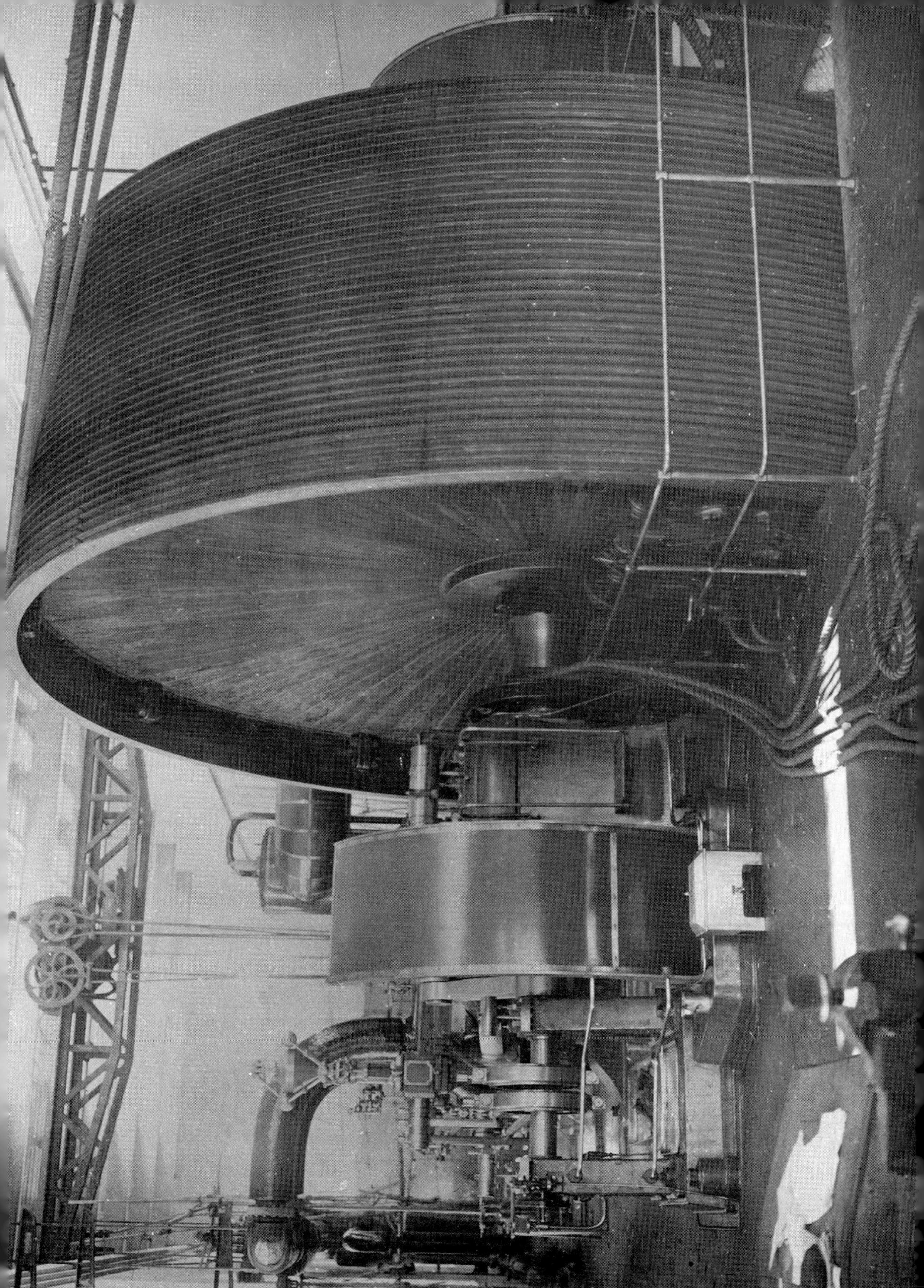

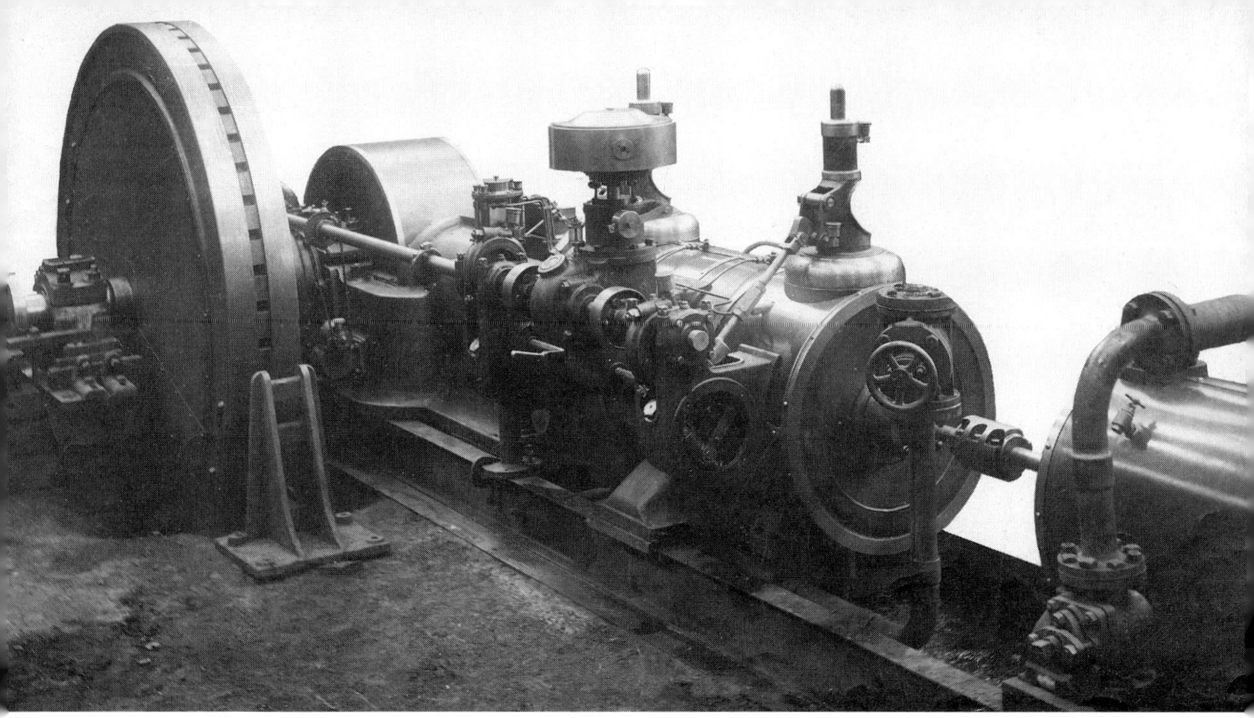

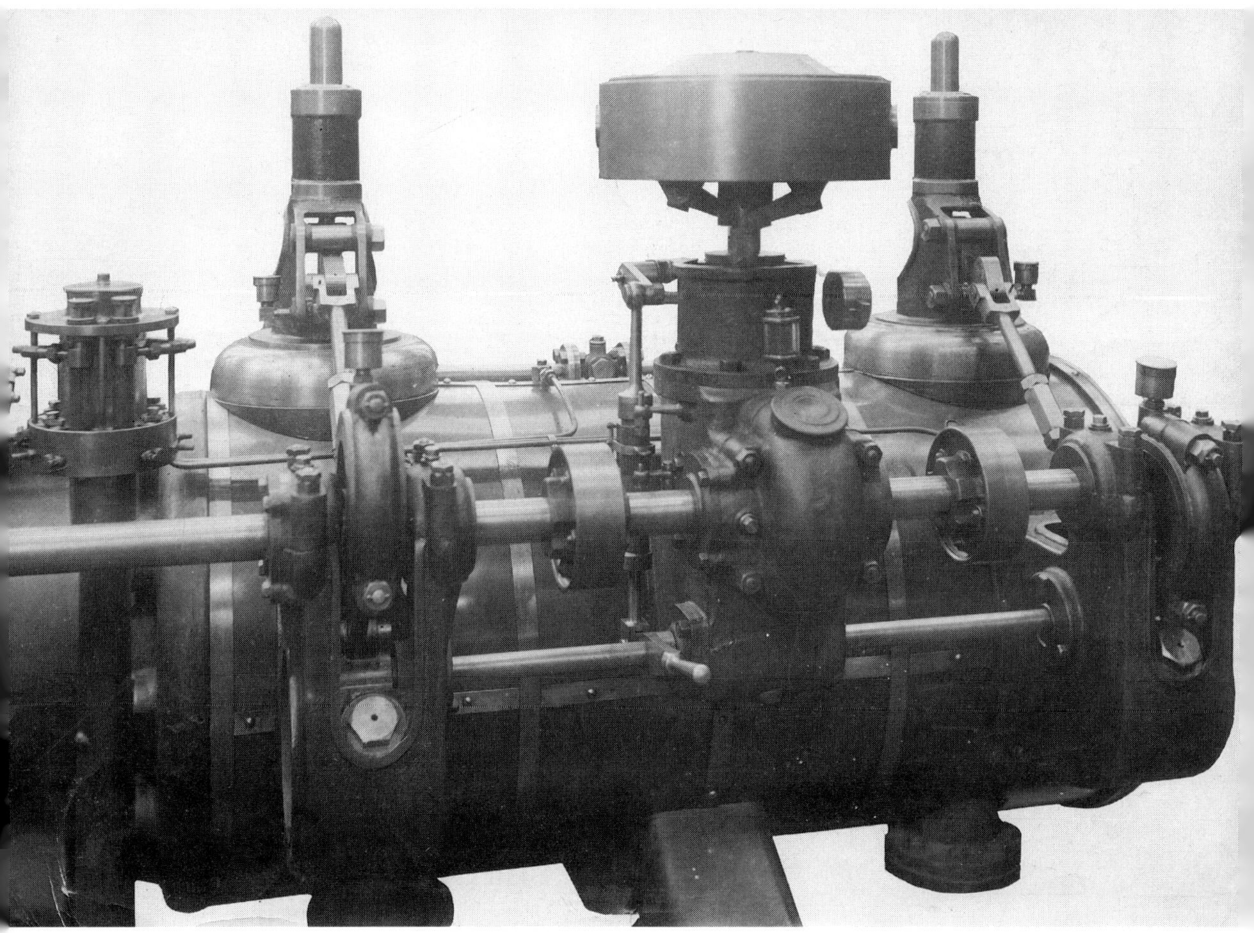

Uniflow engine, 18inch stroke, J. Carmichael Co.

MAK/066 - John Chadwick, Princes Bridge Iron Works, Manchester.

1) Advertisements:
1874 (2).

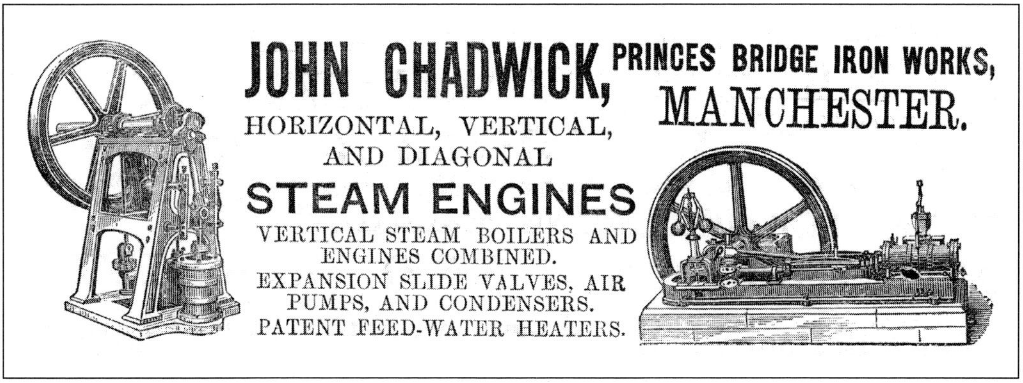

MAK/067 - Chapman & Furneaux, late Black Hawthorn, Newcastle.

1) Advertisements:
Wrapper from *Engineering*, 1 May 1903 with notice of purchase of firm by Hawthorn Leslie.

Reverse: 1903 Marshall & Sons, Gainsborough; Pulsometer Co., Reading; B. S. Massey, Openshaw, Manchester, - pneumatic hammers.

MAK/068 - Chapman & Sutton, Vulcan Iron Works, Rhodeswell Road, Limehouse, London.

1) Advertisements:
1874.

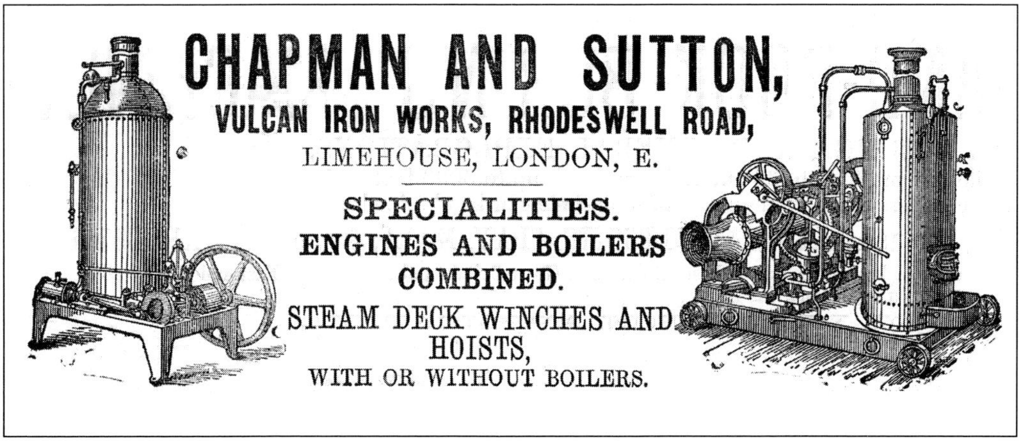

MAK/069 - Charlesworth, Hall & Co., Ltd, Richmond Hill Ironworks, Oldham, Lancs.

1) Advertisements:
1893.

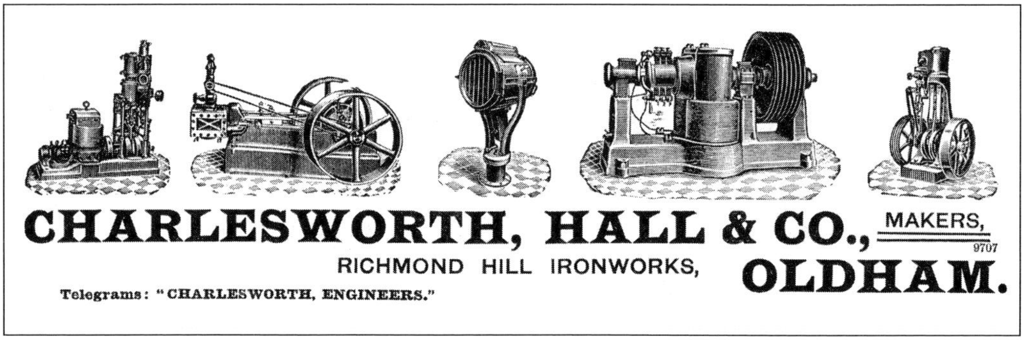

MAK/070 - Clarke, Chapman & Co., Victoria Works, Gateshead upon Tyne.

1) Advertisements:
1893, 1894, 1901, 1912.
Reverse: 1893, Ransome, Sims & Jefferies, Ipswich
Reverse: 1894, Selig, Sonnenthal & Co., 85 Queen Victoria Street and Lambeth Hill, London. - flexible shaft.
Reverse: 1901, R. Hornsby, Grantham. - oil engines
Reverse: 1912 Chesters Engineering, Renfrew, Scotland. - mining machinery.

6) Trade literature:
Booklet: *A few illustrations of our manufactures* n.d., 50pp. [A random collection of leaves from a larger catalogue].

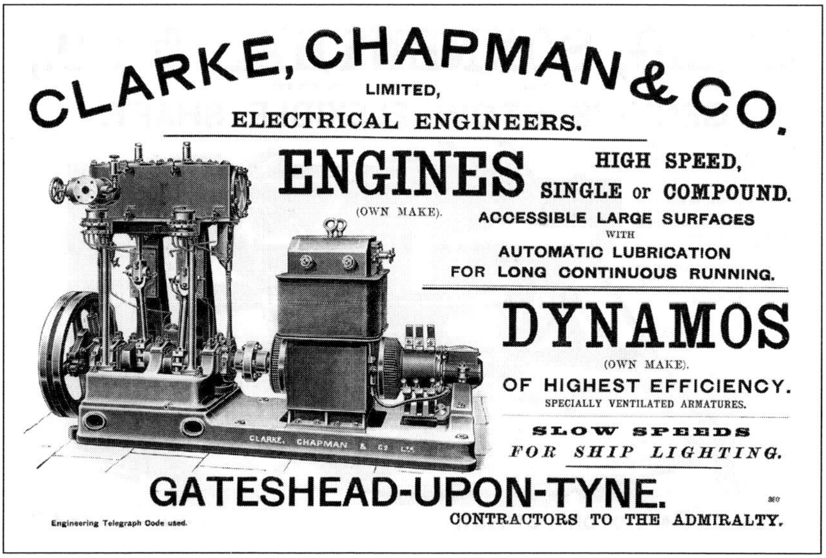

Advert for Twin vertical engine and Dynamo, 1894.

MAK/071 - Joseph Clayton, Soho Foundry, Preston, Lancs.

1) Advertisements:
1885.

2) Letters:
Letter to GW from Harris Public Library, Market Square, Preston, dated 3 March 1970 about Clayton's history.

3) MS material:
MS notes by GW about Claytons.

MAK/072 - Clayton & Goodfellow, Atlas Works, Blackburn.

2) Letters:
Letter to GW from G. A. Lacey, chief draughtsman of Clayton & Goodfellow, dated 2 November 1953.

Letter to GW from Blackburn Public Library, dated 2 January 1974, enclosing letter to them from Clayton and Goodfellow which said all early records had been destroyed.

Carbon copy of letter from GW to Blackburn Public Library, dated 7 January 1974, in reply to 2) above. Says he once saw the works order books and made a transcript of part of it.

3) MS material:
MS notes by GW about the firm. 2pp.

MS transcript by GW from 'Index Book' to works order book, n.d., 4pp.

4) Photographs:
Print of 450 hp vertical Corliss engine at Jas Williamson & Son, Lancaster, scrapped c 1950. (see p.78).

5) Printed material:
MS notes and extracts from a technical periodical about Uniflow designed by J. Davidson.

Pages from *Distaff*, works magazine of Courtaulds Northern Textiles, May/June 1971, with illustrated account of model 500hp horizontal cross compound engine at Royal Mill, Clayton-le-Moors, made by Thurston Smith.

6) Trade literature:
Booklet: *Uniflow steam engine*, n.d. , 16pp.

Part of leaf cut from catalogue, showing horizontal cross compound engine with Corliss gear, n.d.

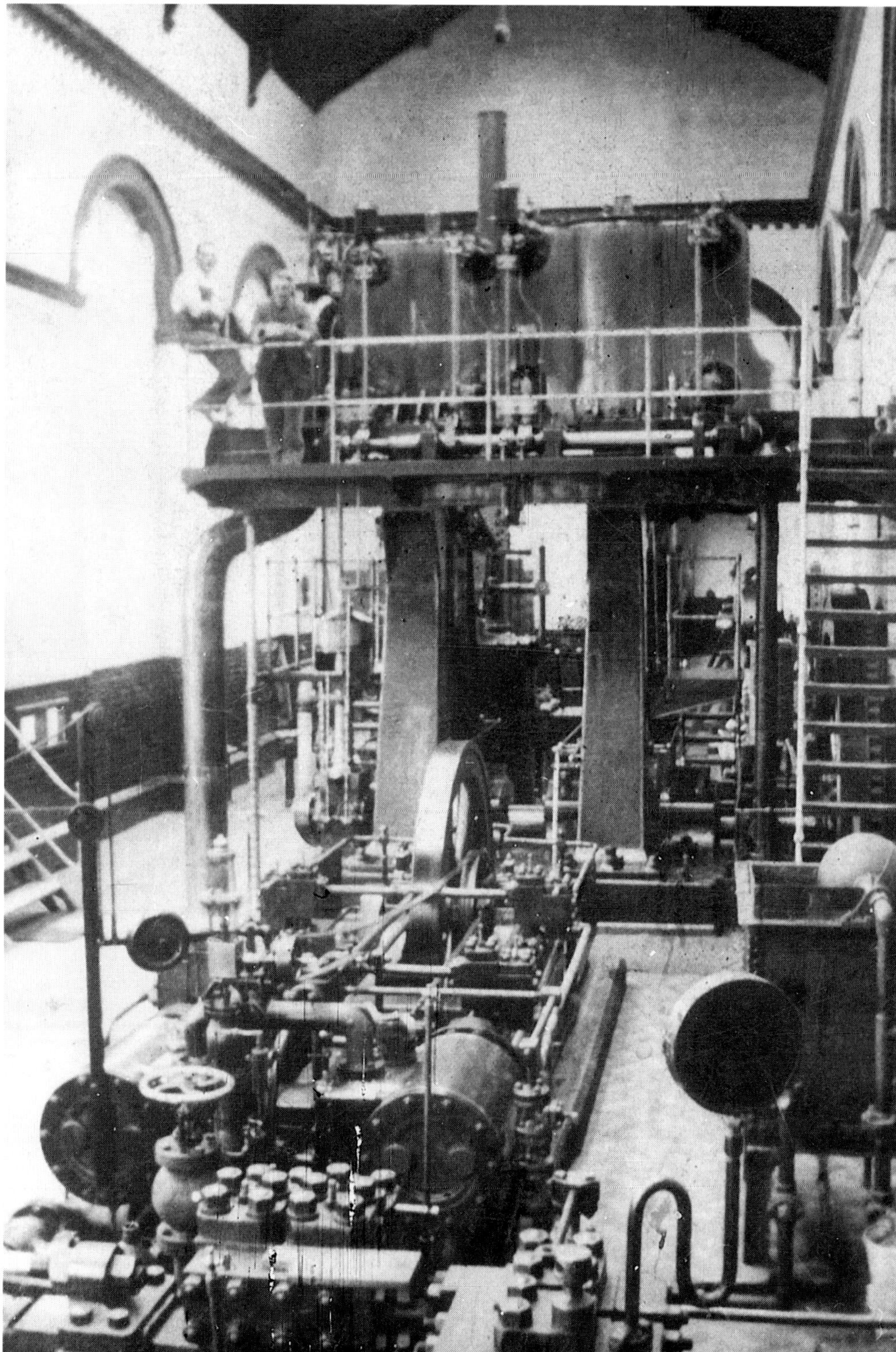

Vertical Corliss Engine at J. Williamson & Son, Lancaster. Clayton & Goodfellow's.

MAK/073 - Clayton, Howlett & Venables, Atlas Works, Woodfield Road, Harrow Road, London W.

1) Advertisements:

1888.

Reverse: 1888 William Johnson, Castleton Foundry, Armley Road, Leeds, - brick-making machinery.

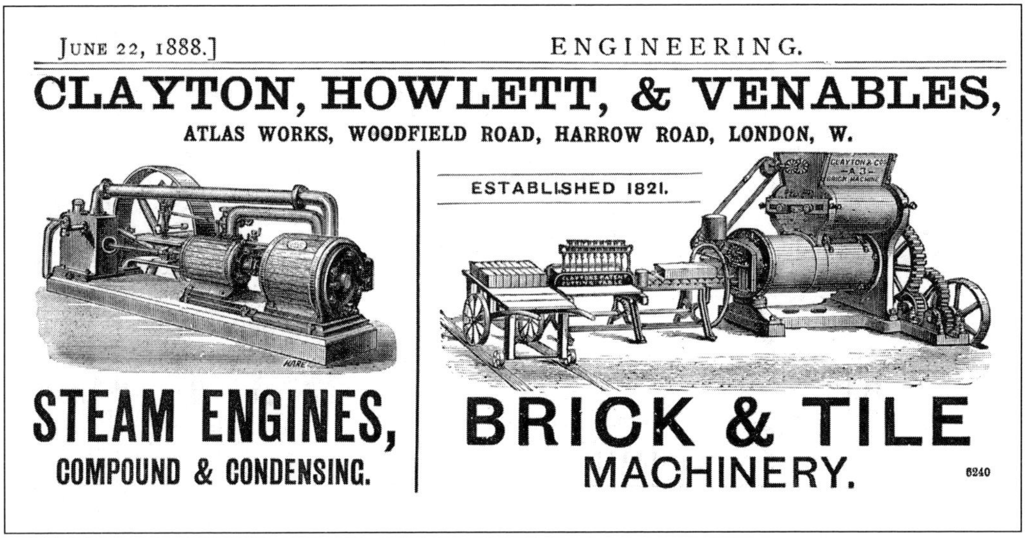

MAK/074 - Clayton & Shuttleworth, Stamp End Works, Lincoln.

1) Advertisements:

1887 (see below), 1894, 1895 Engineering, 8th November, (see p.80).

Reverse: 1895 La Meuse, Liège, Belgium – various engines.

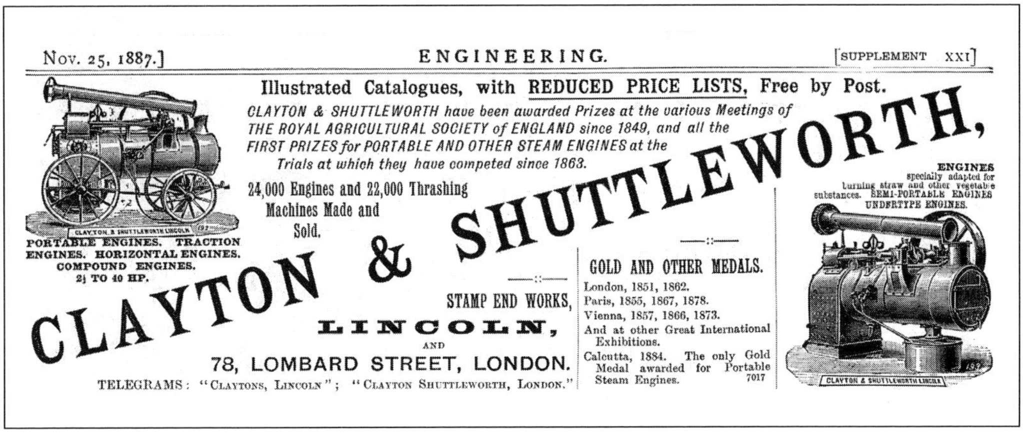

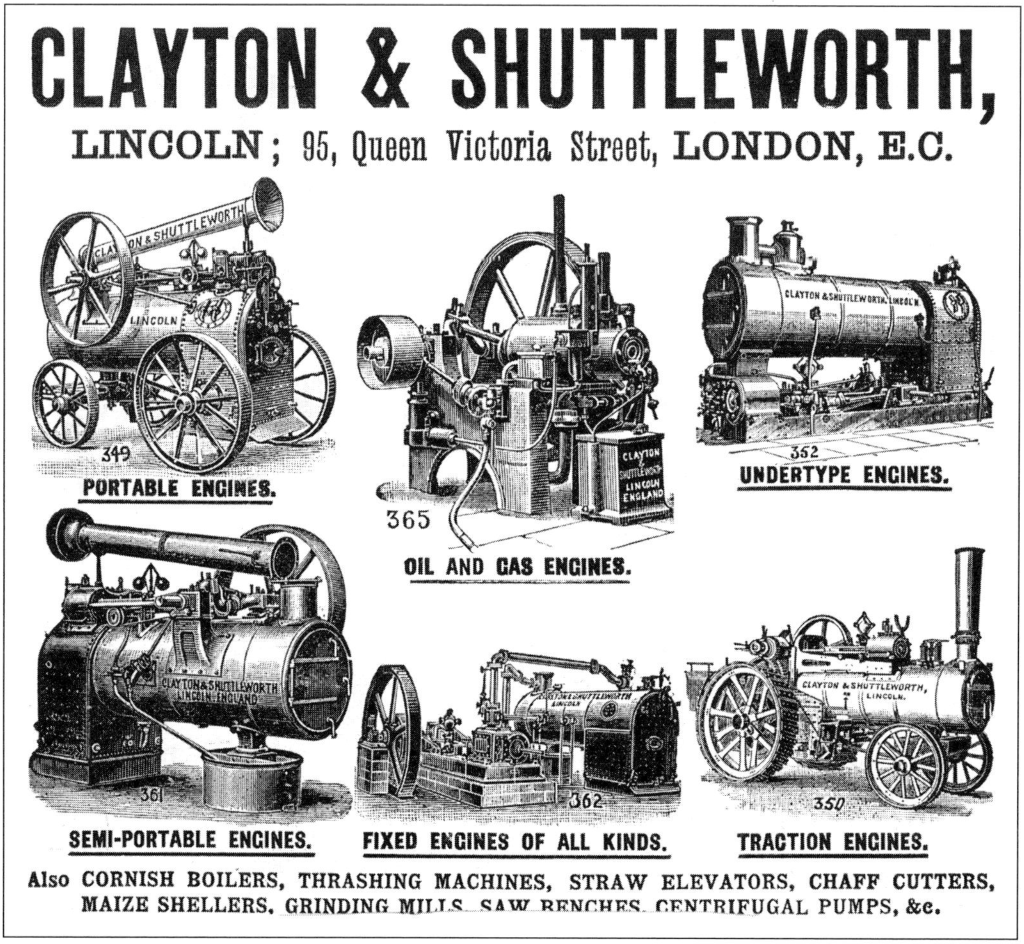

MAK/075 - Coalbrookdale Co.,

1) Advertisements:

1887 (see below).

Reverse: 1887 Hathorn Davey & Co., Leeds. - pumping machinery; F. D. Bumstead, Cannock Chase Foundry, Hednesford, Stafford. - Chandler's 'Silent engine'; M. Immisch, Maiden Works, Prince of Wales Road, London NW. - Electric Motors.

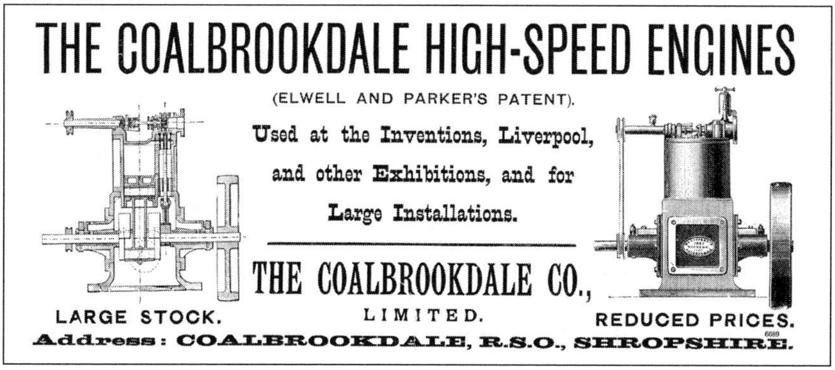

MAK/076 - Victor Coates, Belfast.

5) Printed material:
2 copies of article from *Engineering*, 27 November 1903, about 3000 hp triple expansion engine for Newcastle Corporation Tramways. (see below).

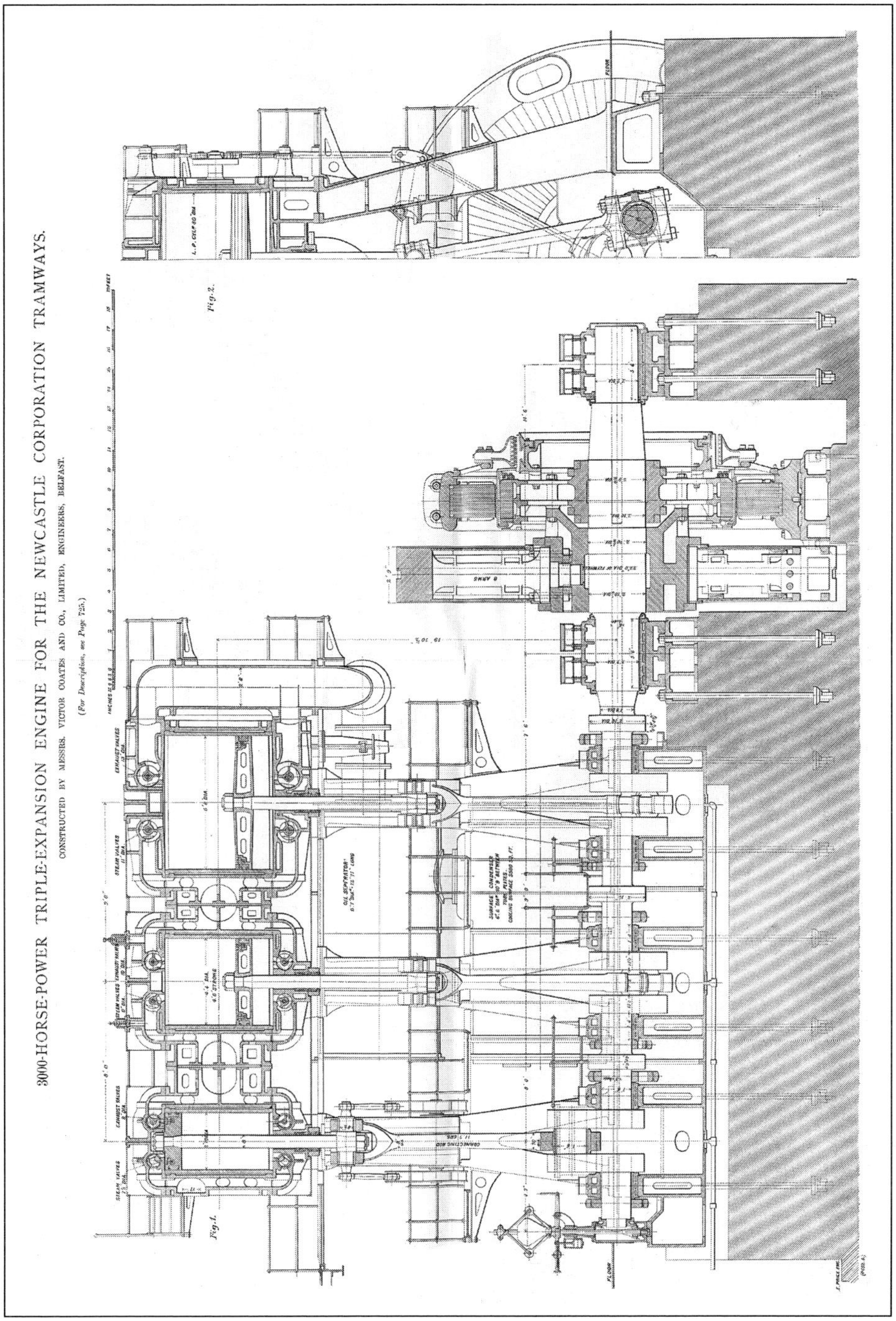

MAK/077 - Cochran, Annan, Dumfriesshire.

1) Advertisements:
1879, 1885, 1887 (see below), 1890 (2).

Reverse: 1879 Patent Steam Boiler Co., Heneage Street, Birmingham. - 'Roots' tubulous steam boiler; H. Titley & Son, Alton Works, Soho, Smethwick.- boilers; Clarke's Crank Co., Patent Crank Works, Lincoln.

2) Letters:
Letter to GW from Cochran Boilers, 30 December 1954.

Letter to GW from Cochran Boilers, 4 January 1955.

5) Printed material:
Photocopy of page from *The Marine Engineer* dated 1 November 1882. engravings of Cochran vertical multi-tubular boiler and engine.

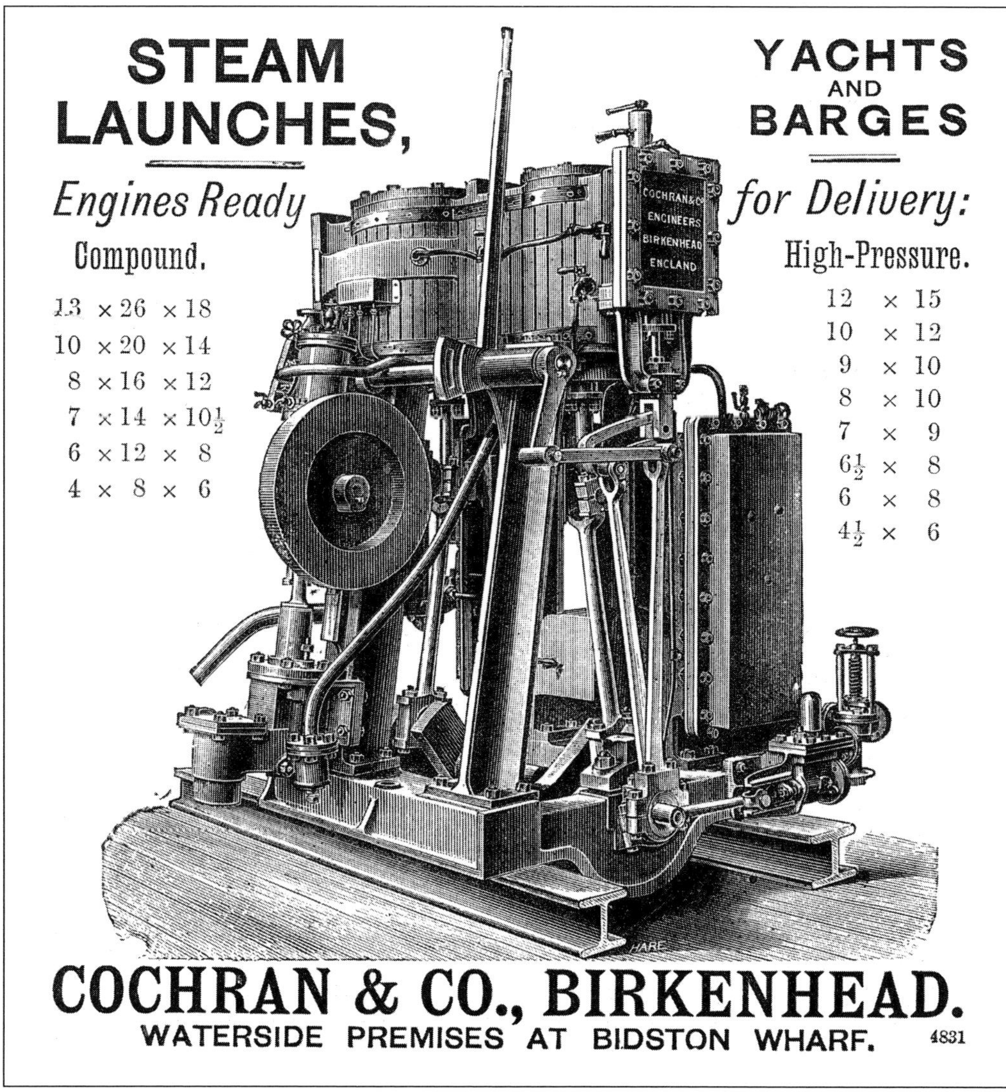

MAK/078 - John Cochrane, Barrhead, by Glasgow.

1) Advertisements:
1886, 1895, 1897 (see below), 1902.

5) Printed material:
Article from *Mechanical World*, 24 November 1888 about Cochrane's exhibit at the Glasgow Exhibition, Horizontal Corliss engine and vertical engine and centrifugal pump.

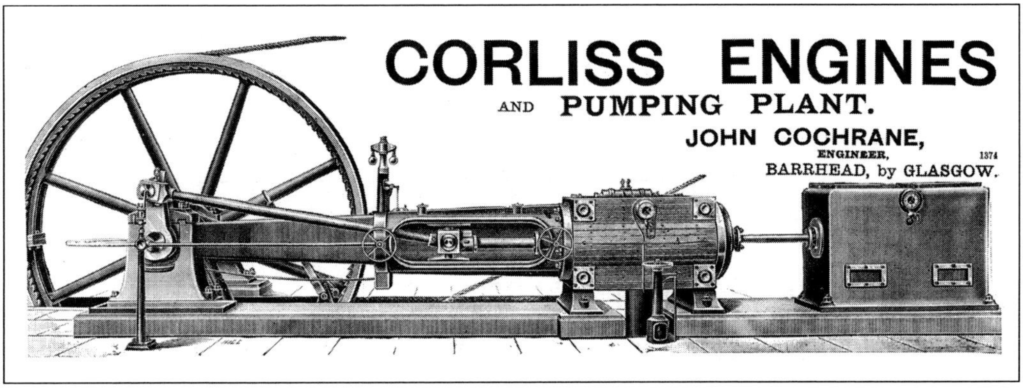

MAK/079 - Tom A. Cockin, Waterloo Mills, Leeds Road, Huddersfield.

5) Printed material:
Photocopy of page 79 from *Industries of Yorkshire*, part 2, 1890. with brief details of the firm.

MAK/080 - Cole, Marchant & Morley, Prospect Foundry, Bradford, Yorks.

2) Letters:
Letter to GW from David & Charles, 24 September 1970, with his reply dated 28 September 1970 about advertising booklets he had been sent.

Letter to GW from Arnold Throp, dated 28 March 1979. Discusses some of the advertising booklets noted below.

3) MS material:
Photocopy of transcript of Cole, Marchant & Morley's engine production, 1895-1925. 32pp Origin unknown.

5) Printed material:
Article from *Engineering*, 15 March, 1889 about triple expansion engine at Hinchcliffe's Mills, Cragg Valley, Mytholmroyd. (see pp.85-86).

Article from *Mechanical World*, 18 November 1892, about 400 hp horizontal mill engine for a spinning mill at Laisterdyke. (see p.84).

Photocopy of article from *Engineering*, 26 May 1893 about horizontal compound engine with Morley's positive valve gear.

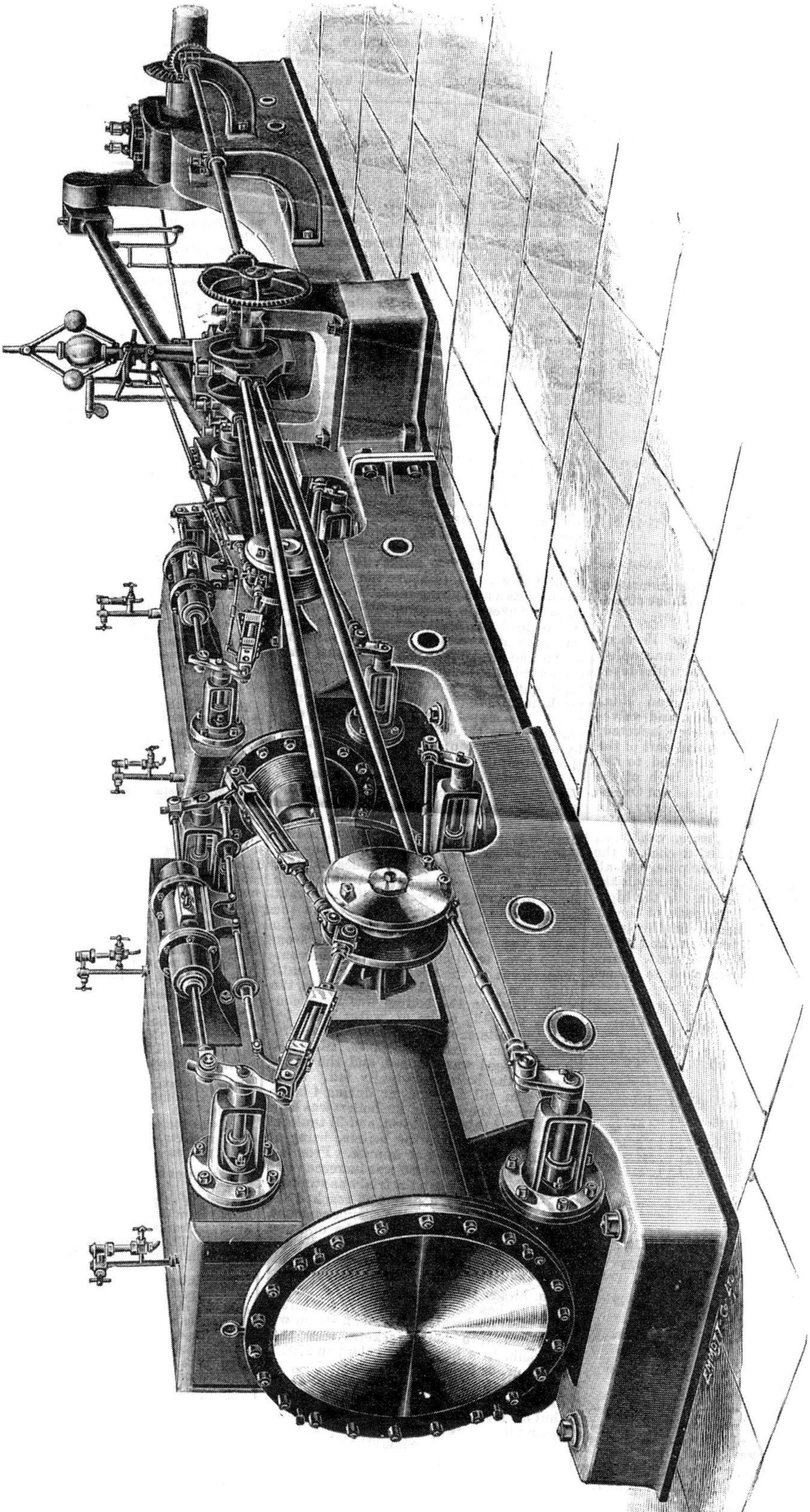

400I.H.P. COMPOUND MILL ENGINE

TRIPLE-EXPANSION ENGINE AT MESSRS. R. AND H. HINCHCLIFFE'S MILLS

CRAGG VALLEY, MYTHOLMROYD.

CONSTRUCTED BY MESSRS. COLE, MARCHANT, AND CO., ENGINEERS, BOWLING, FROM THE DESIGNS OF MESSRS. RHODES AND CRITCHLEY, ENGINEERS, BRADFORD.

(*For Description, see Page* 264.)

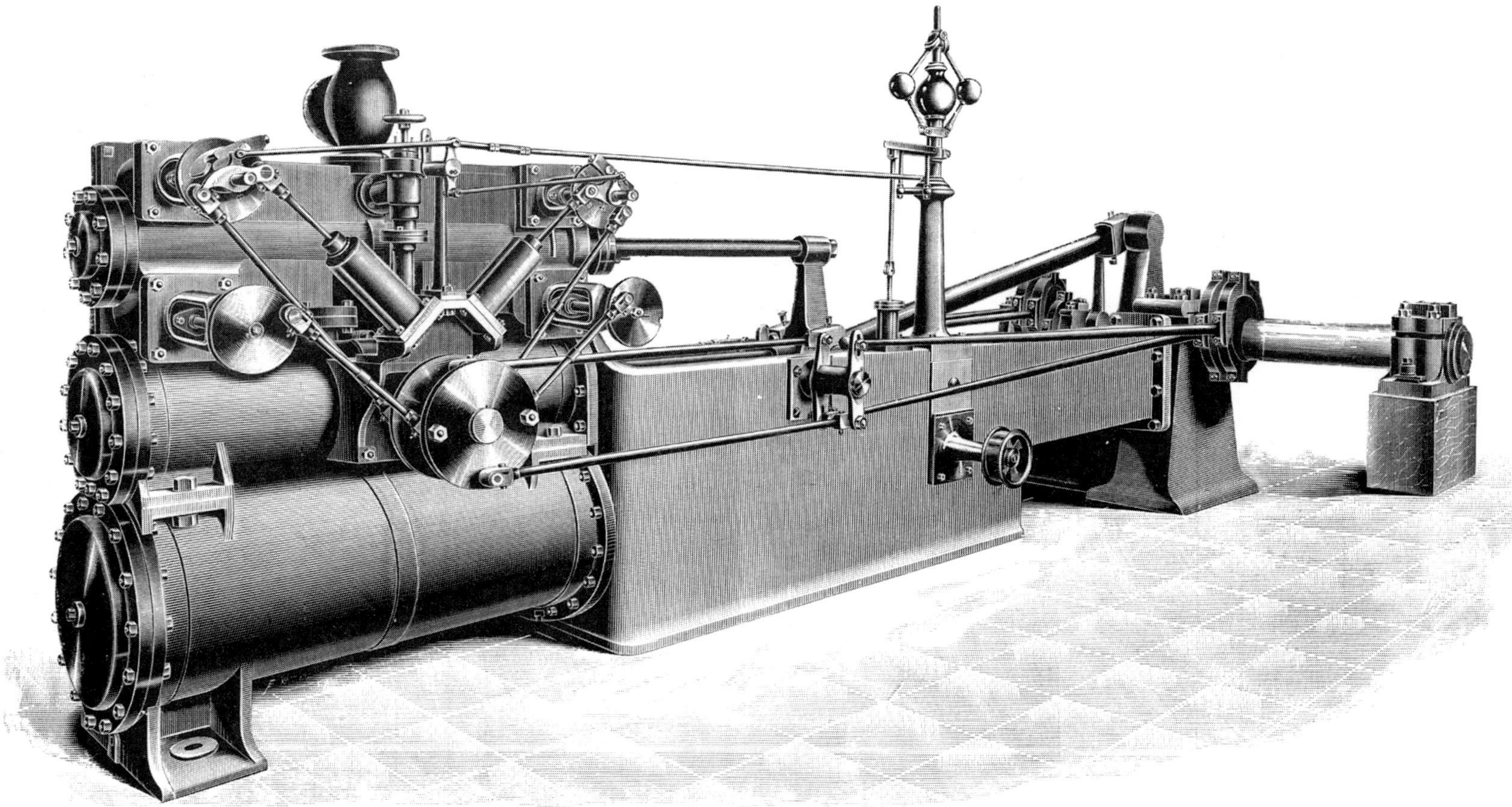

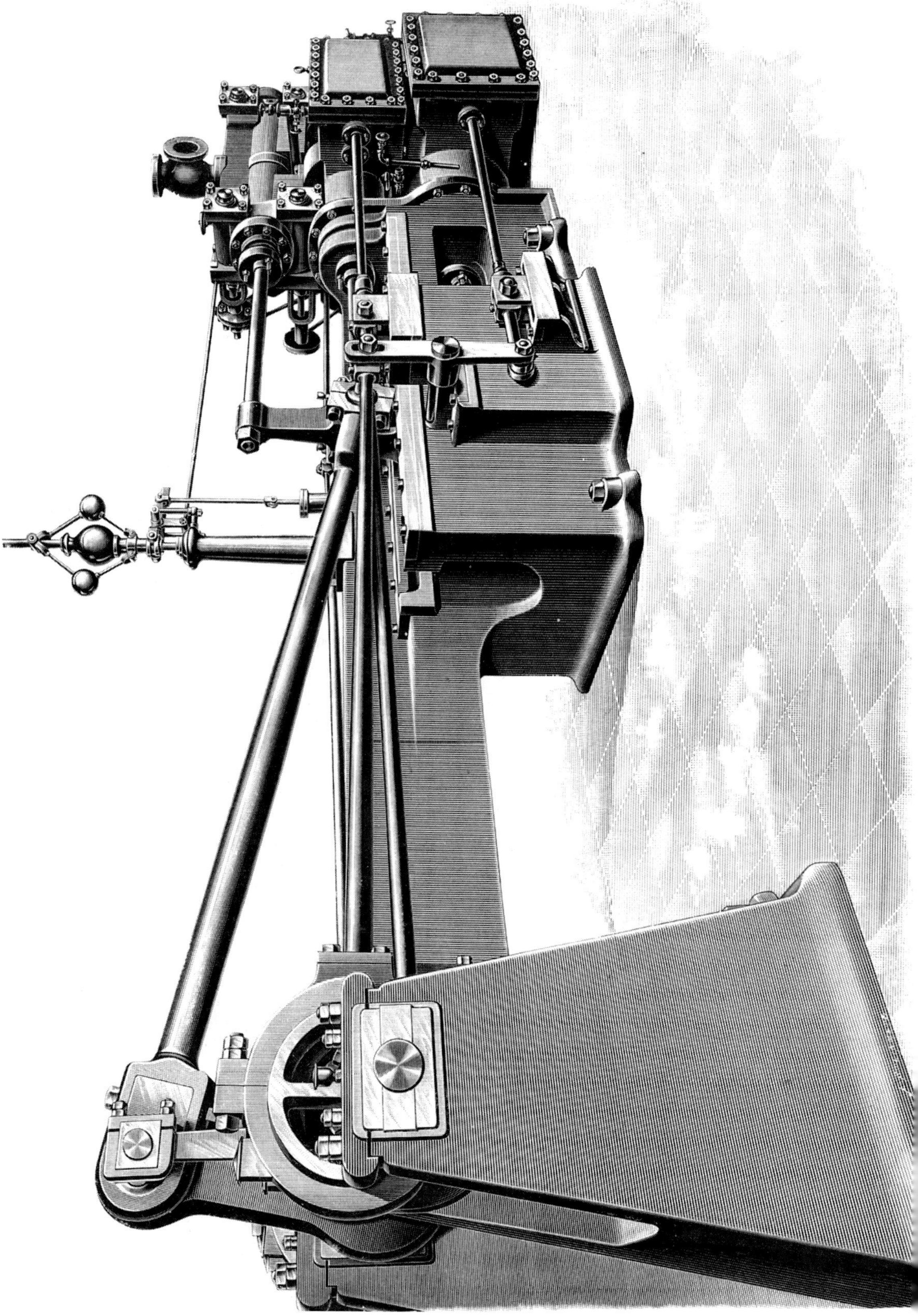

MAK/081 - H. Coltman & Son, Loughborough.

1) Advertisements:
1882, 1886 (2), 1887, 1888, 1910.

From 'Engineering'.

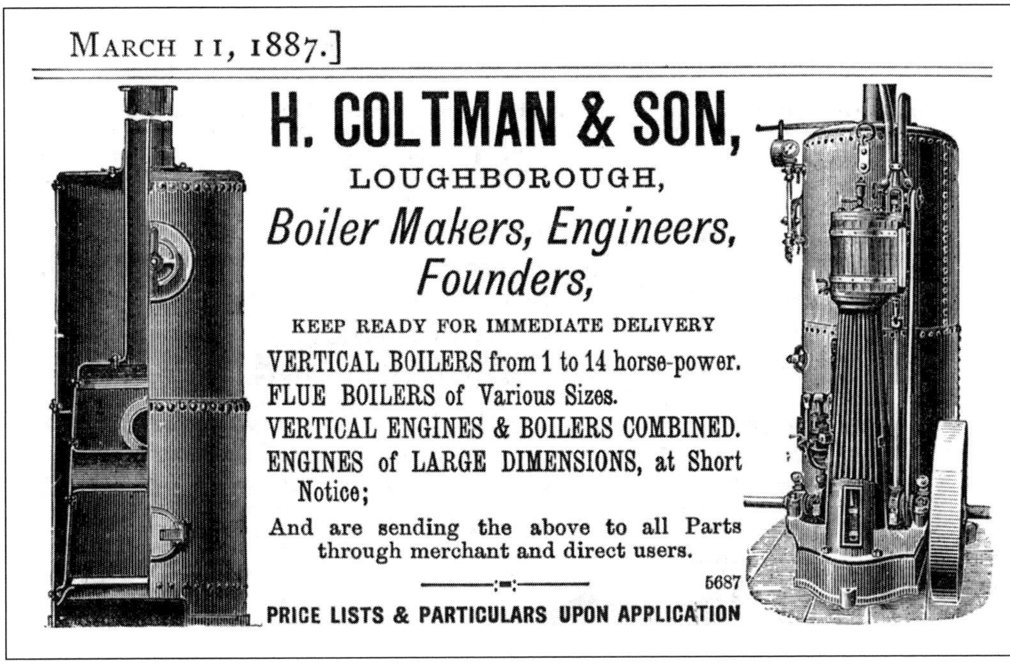

MAK/082 - Combe, Barbour & Combe, Falls Foundry, Belfast.

1) Advertisements:
1879 [Block illustrating the works].

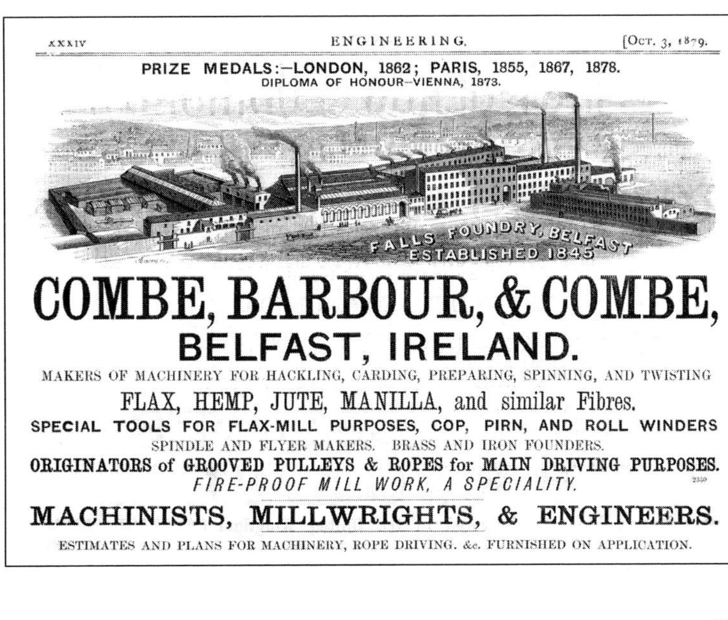

MAK/083 - Compagnie Des Fives-Lille, Lille, France.

5) Printed material:
Page 632 from *Engineering* 16 Nov. 1900 with 2 pix of compound engine and 3-phase alternator at the Paris Exhibition. Descriptive text (p 626) missing.

MAK/084 - Corliss, USA.

5) Printed material:
Article from *Engineering*, 19 May 1876, p 413 describing the 1400 nhp Corliss beam engine at the Philadelphia Exhibition, with full page plates. Mentions an earlier part of the description at p 331 (missing). (see below & opposite).

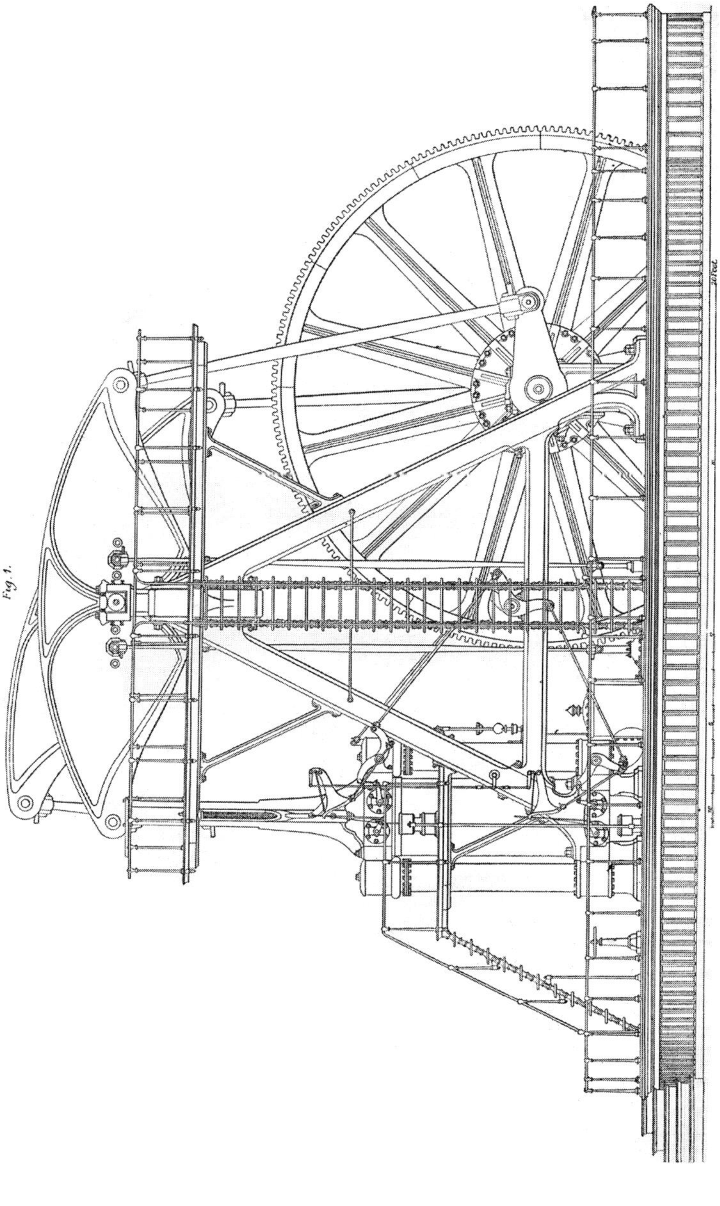

CORLISS BEAM ENGINE (1400 HP. NOMINAL) AT THE PHILADELPHIA EXHIBITION.

CONSTRUCTED BY MESSRS. GEORGE H. CORLISS AND CO., ENGINEERS, PROVIDENCE, RHODE ISLAND, U.S.A.

(For Description, see Page 413.)

Fig. 1.

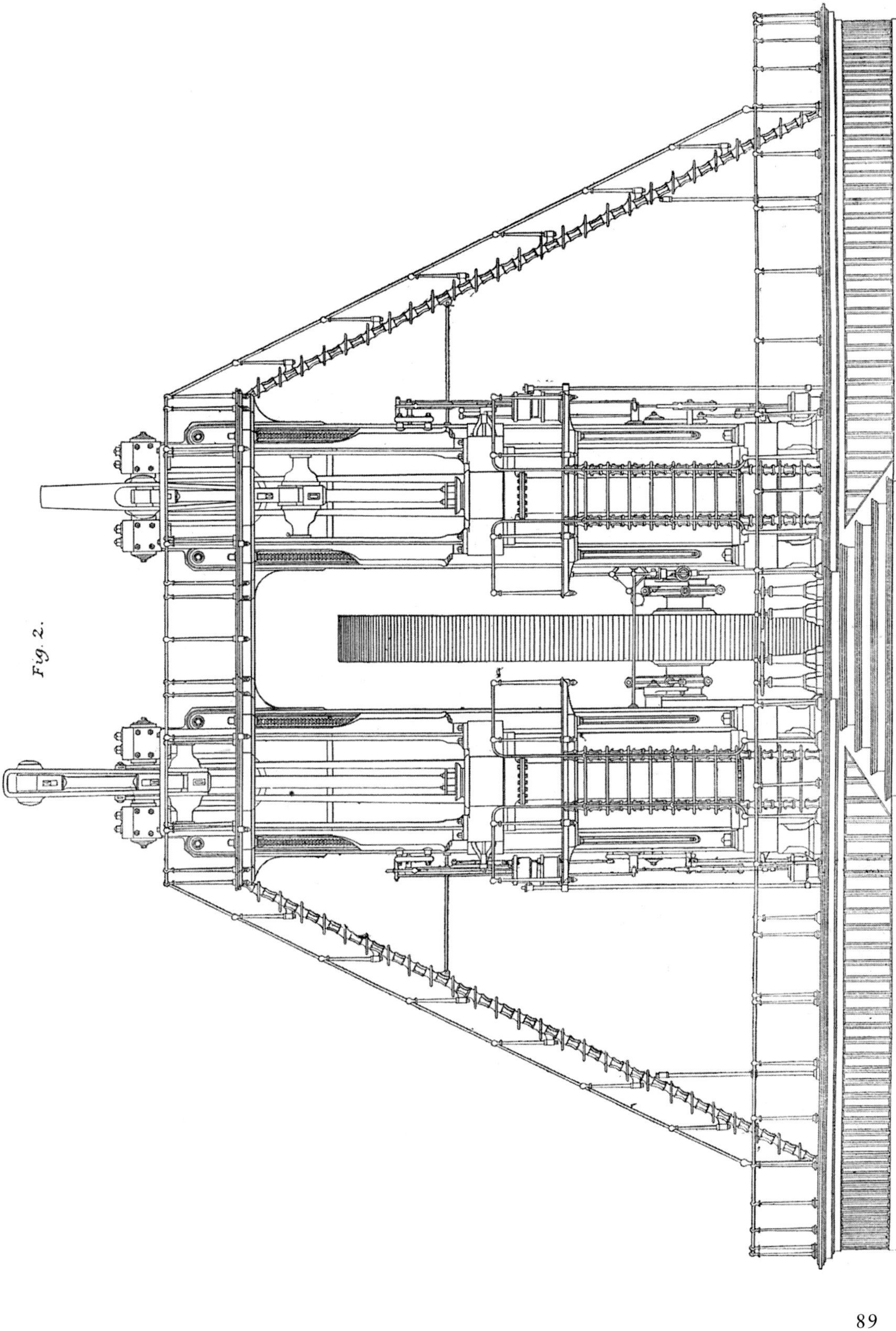

Fig. 2.

MAK/085 - R. J. & E. Coupe, Worsley Mesnes Iron Works, Wigan.

1) Advertisements:
1886 advertising the firm for sale on the death of the senior partner.

MAK/086 - A. F. Craig & Co., Ltd., Caledonia Engine Works, Paisley, Scotland.

1) Advertisements:
c 1900.

MAK/087 - Crane Packing, Ltd., Slough, Bucks.

6) Trade Literature:
Booklet:, *Metallic Packing*, 5 cd, 124 pages, n.d. , but c 1920.

MAK/088 - Crompton & Company Ltd, Salisbury House, London E.C.

1) Advertisements:
1905, 1907.

Reverse: 1907 Phoenix Bolt & Nut Co., Handsworth, Birmingham; Cedervall - protective lubricating box for propellor shafts; Ingersoll-Rand - electro-pneumatic air drill.

MAK/089 - Thomas T. Crook, Stanley Ironworks, Derby Street, Bolton, Lancs.

6) Trade literature:
Collection of 12 printed leaflets.

1) 9ft edge runner mill.
2) *Stanley* and *Perfect* edge runner mills.
3) Contractor's steam winch.
4) 10 cwt petrol friction hoist.
5) Horizontal pug mill.
6) Plastic brick-making machine.
7) Patent concrete and tarmac mixers.
8) Horizontal Corliss engine.
9) Horizontal high-pressure engine. (see p.91).
10) Horizontal high-pressure engine.
11) Hand lever brick and tile press.
12) Complete brick making plant with detailed technical description.

ESTABLISHED 1864.

THOMAS T. CROOK,

STANLEY IRONWORKS, BOLTON, Lancashire.

Engineer, Millwright, Iron and Brass Founder.

MANUFACTURER OF
VERTICAL, HORIZONTAL, AND DIAGONAL STEAM ENGINES, CONDENSING AND NON-CONDENSING.
GENERAL MILLWRIGHT WORK, BEVEL, SPUR AND ROPE DRIVING.
STEAM WINCHES AND CONTRACTORS' PLANT.
IMPROVED PLASTIC BRICKMAKING MACHINERY, suitable for either Shale or Clay.
MORTAR MILLS, WITH ENGINES COMBINED, OR PULLEYS (under or over Driven), with 5-ft. to 9-ft. Pans.
FEED WATER HEATERS.
PATENT EDGE RUNNER GRINDING MILLS. MORTAR MILLS ALWAYS KEPT IN STOCK.

Horizontal High-Pressure Engine,

WITH AUTOMATIC EXPANSION GEAR.

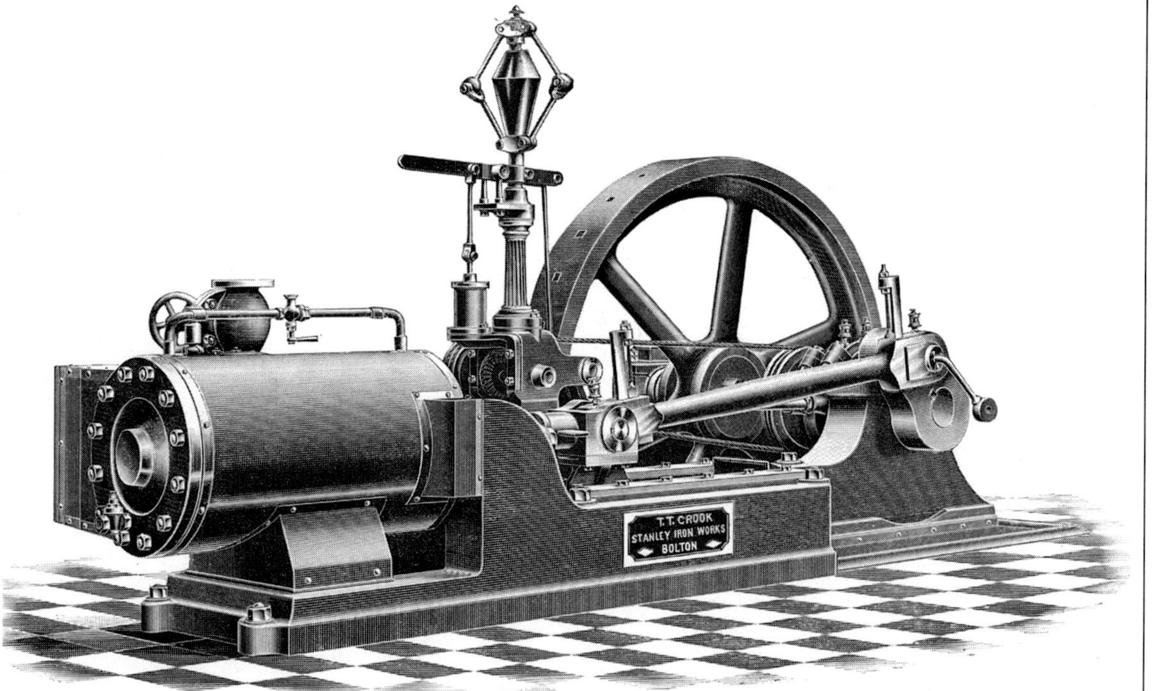

THE above illustration represents my Horizontal Engine fitted with Automatic Expansion Gear. The construction of this Engine is the same as described on Sheet 2, with the exception that Automatic Expansion Gear is fitted. This gear is designed to economise the consumption of steam in the simplest possible manner. The gear is of the link type, controlled by a very sensitive and powerful high-speed Porter Governor, and capable of regulating the Engine to one required speed, and utilizing the steam to the greatest advantage. With this gear the equilibrium valve is dispensed with, and the Engine is governed without "wire-drawing" or reducing the pressure of steam supply. The steam is admitted to the cylinder at boiler pressure up to the point of cut off, and is cut off sharply at any point varying with the load on the Engine. The results obtained from the working of this type of Engine compare very favourably with Corliss and other intricate valve gear Engines, and has the advantage of being less costly, simpler, and requiring much less attention, repair, and adjustment than any other gear. All the materials used in the construction are the best, and are well finished throughout. Horizontal or Vertical Air Pumps and Condensers can be fitted if required.

MAK/090 - Dake, USA.

6) Trade literature:

1) Booklet: *Instruction book … Dake engines … repair parts list* 30pp, n.d. but c 1920.

MAK/091 - Davey Paxman, Colchester, Essex.

1) Advertisements:

1887 (see below top), 1891, 1895, 1896, 1896, 1903, 1905, 1907, 1909, 1909, 1910, 1910, 1911.

Reverse, 1895, Ebbw Vale Steel & Iron and Coal Co.,
Reverse 1905, 1907 J. Halden - drawing instruments etc.
Reverse 1909, Fielding & Platt, Gloucester.

5) Printed material:

Photocopy of p 346-7 from Haeder & Powles, *A Handbook on the Steam Engine*, 1896.

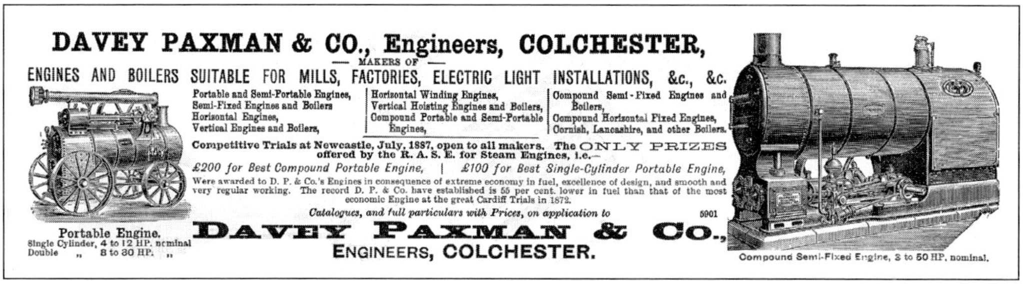

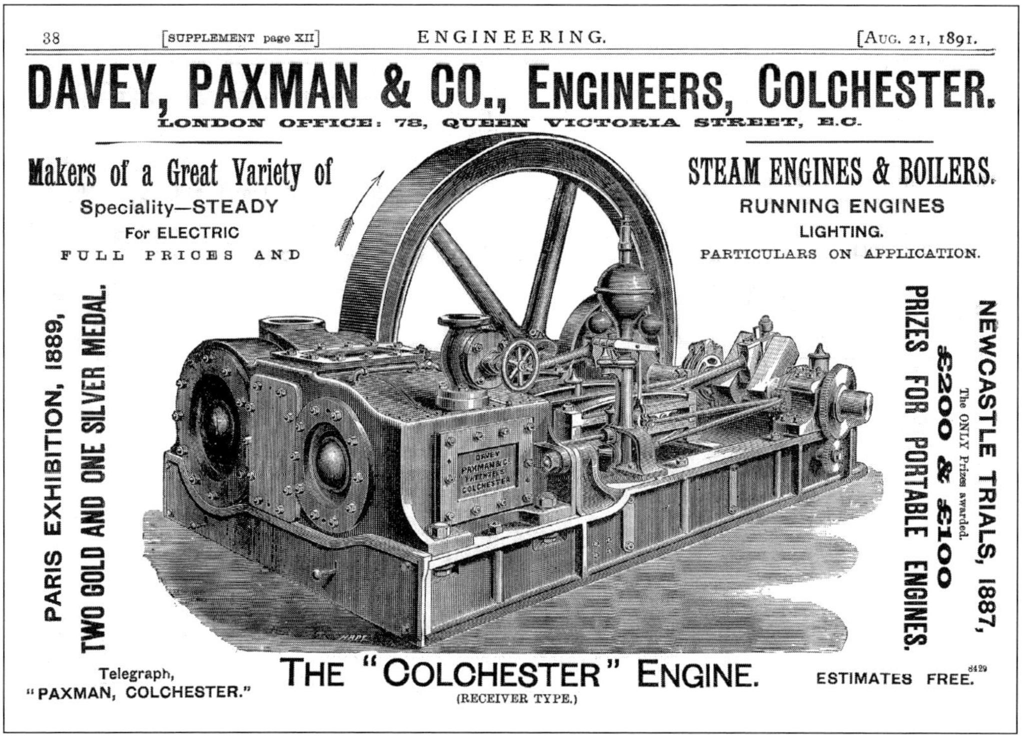

MAK92- Davidson, Manchester.

5) Printed material:
Photocopy of plate from F. Dye, *Popular Engineering*, n.d. [add details].

MAK/093 - Davy Bros, Park Ironworks, Sheffield.

1) Advertisements:
c 1865.

3) MS material:
Sheet of MS notes by GW. Note: the address is given Damell Works, Sheffield. Says most engine drawings were lost in the Blitz. Firm made last steam engine in 1913.

Sheet of MS notes by GW. Appears to be summary of information from *Engineering,* 6 April 1883, p 320.

5) Printed material:
Pages extracted from *Engineering*, 26 December 1902, illustrating a 300 kW vertical TE engine & dynamo for Middlesbrough Electricity Co.,

6) Trade literature:
Card from United States Metallic Packing Co., n.d. but pre-1914, illustrating Davy Bros 16,000 hp 3-cyl high pressure vertical reversing engine for Dorman Long's Britannia Ironworks, Middlesbrough. (right).

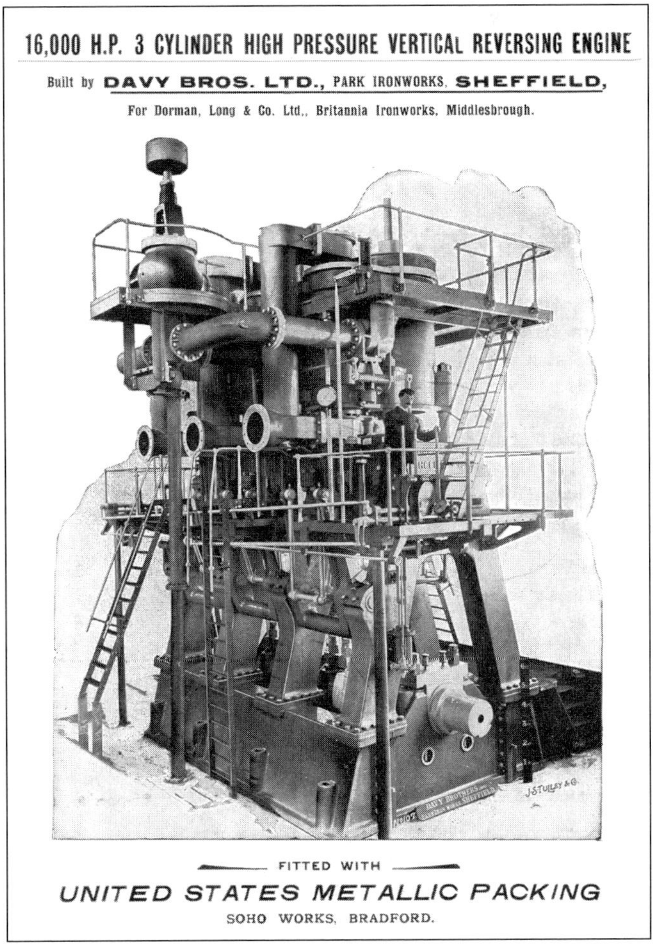

MAK/094 - Day & English, Engineers, Bath.

1) Advertisements:
1887.
Reverse: 1887 John Fowler, Leeds.

MAK/095 - Day, Summers & Co., Northam Iron Works, Southampton.

1) Advertisements:
1901.
Reverse: 1901 Crossley Bros, Manchester - Crossley's gas dynamos.

MAK/096 - Delaunay-Belleville Co., Paris.

5) Printed material:
Article from *Engineering*, 13 July 1900 about 1250hp engine at the Paris Exhibition.

MAK/097 - Des Vignes, Upper Thames.

1) Advertisements:
1905 Sale of the works.

MAK/098 - Dewhurst, Taylor, Heywood, Lancs.

2) Letters:
Letter dated 26 March 1976 to GW from Riley & Son. about the history and about several engines they made.
Letter dated 13 June 1976 from GW to Riley's in reply with valuable comments about his problems in getting his books published.

MAK/099 - Dingler Maschinenfabrik A. G. of Zweibrücken.

5) Printed material:
Article from *Engineering* 26 September 1902 about a 600 HP vertical compound engine at the Düsseldorf exhibition.

MAK/100 - Alfred Dodman, Highgate Works, King's Lynn.

1) Advertisements:
1895 (right), 1903 (below).

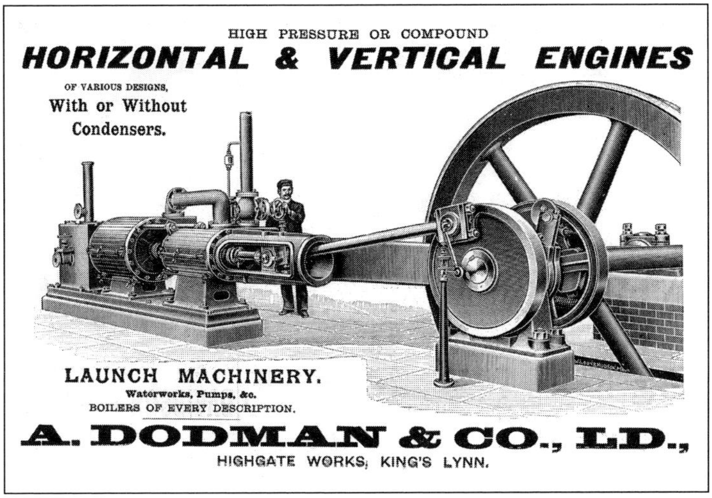

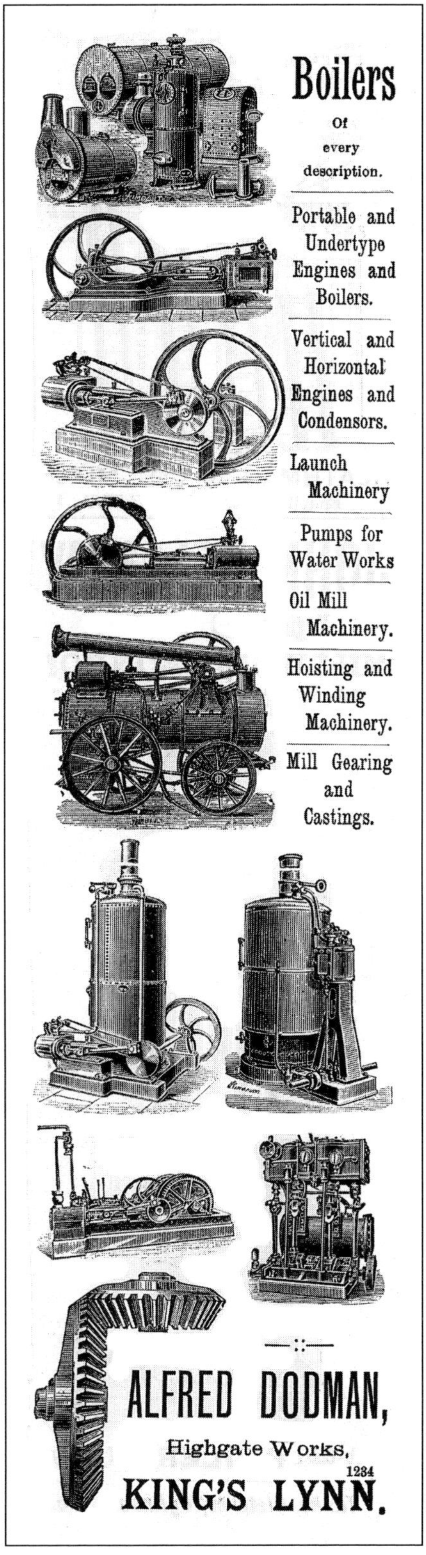

MAK/101 - Bryan Donkin Co., Ltd, Chesterfield.

1) Advertisements:
1893 (see below), 1901(2), 1903, 1903, 1906 (see opposite).

***2) Letters*:**
Letter dated 5 February 1959 from Donkins with 1) list of engines supplied to J. S. Fry & Co., - mentions having list of all engines supplied; 2) mentions recent reorganisation when 'large amount of old data and photographs were recently destroyed' but firm had retained some detail linen-backed drawings.

3) MS material:
MS note by GW about the B. W. Farey patent no 425 of 1866.

4) Photographs:
Print of the model Donkin-Farey engine based on that at Fry's, now in Bristol Industrial Museum. (see p. 98).

5) Printed material:
Article from *Engineering* 3 November 1871 describing tests on the Donkin-Farey engine at Hele works [papermill] Cullompton, Devon.
Article from *Engineering*, 24 April 1874 describing a horizontal compound engine.

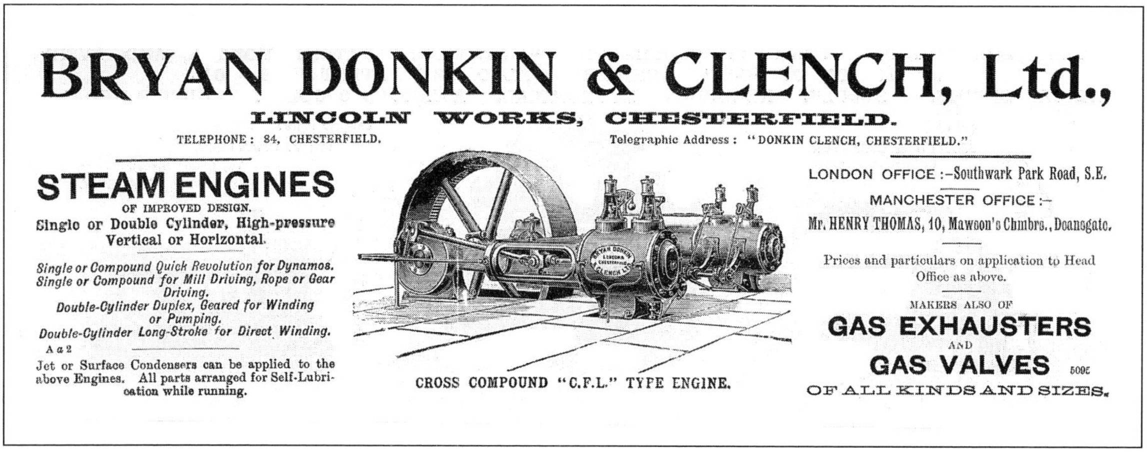

The Bryan Donkin Company, Ltd.,

CHESTERFIELD.

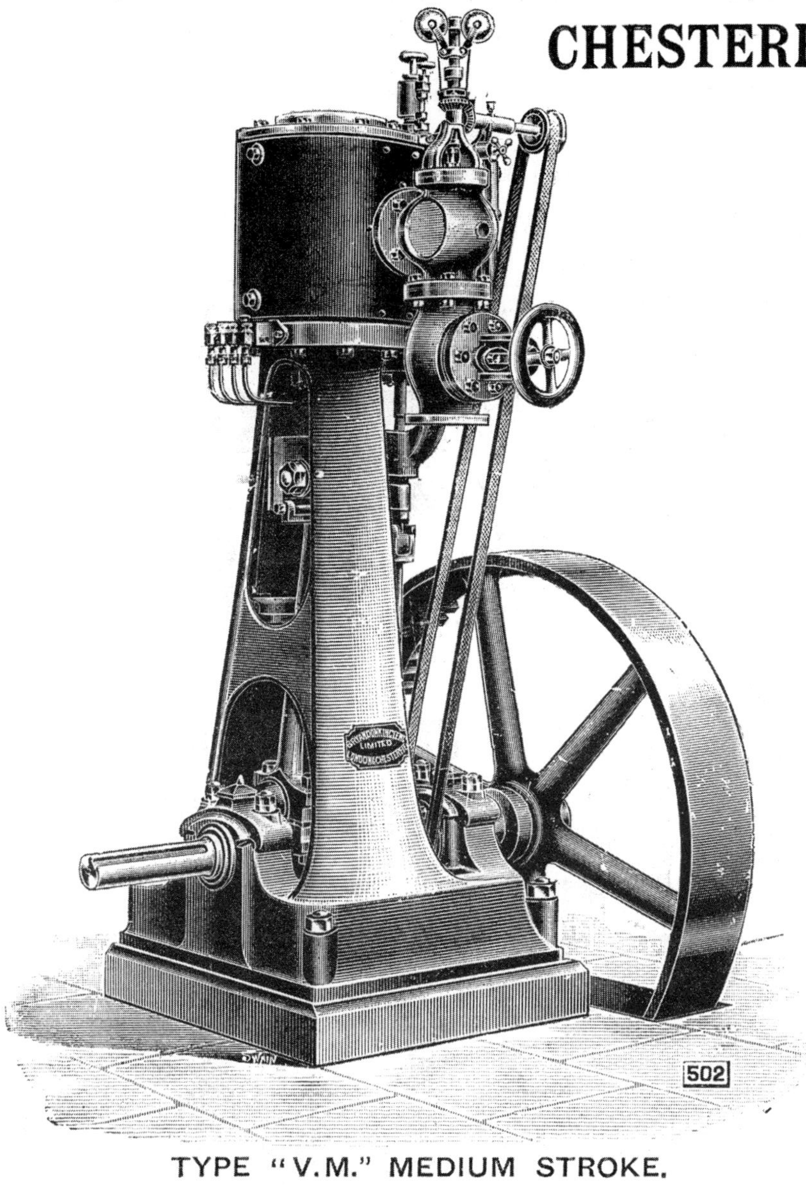

TYPE "V.M." MEDIUM STROKE. 7527

ALSO MAKERS OF ALL OTHER TYPES OF STEAM ENGINES.

Valves of all kinds and Gas Exhausting Plants a Specialty.

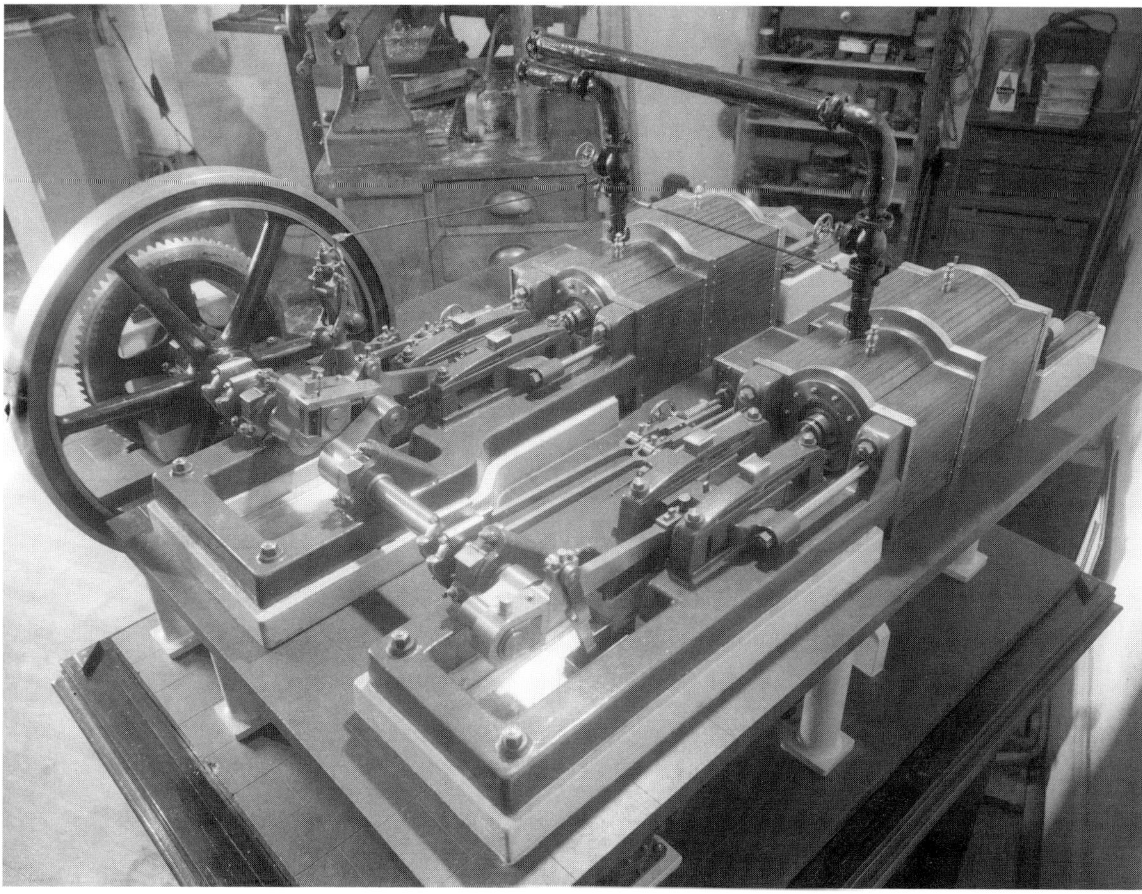

Model of a Donkin Farey engine (see p. 96)

MAK/102 - Douglas & Grant, Dunniker Foundry, Kirkcaldy.

1) Advertisements:
c 1900, 1903, 1908, 1909, 1914.
Reverse: 1914, Penman & Co., Glasgow - boilers.

2) Letters:
Letter dated 12 August 1966 to GW from Strahan & Livingston, publishers of *Fife Mail* about the firm's history.

Letter dated 1 Sepember 1966 to GW from Kirkcaldy Public Library about the firm's records.

Letter dated 12 July 1967 to GW from Lewis C. Grant Ltd about the firm's history.

4) Photographs:
Photograph of engraved catalogue: illustration. No detail noted.

5) Printed material:
Article and plate from *Engineering*, 19 May 1876, about horizontal Corliss engine. (see below).

Photocopy of plate from *The Engineer*, 26 January 1877 of compound beam engine, with MS note by GW.

Plates from *Engineering*, 6 April 1888, of horizontal 2500hp compound Corliss engine.

Article from *Mechanical World*, 28 December 1889 and 4 January 1890 about vertical triple expansion inverted cylinder engines.

Plate from *Engineering*, 30 December 1910 of 600 IHB compound with Carels balanced drop valves.

7) Other material:
TS *Messrs Douglas & Grant Ltd., Dunniker Foundry, Kirkcaldy*, 7pp no date, no author.

TS *Historical notes regarding the Business of Douglas and Grant Ltd, Kirkcaldy & Lewis C. Grant Ltd, Dysart, Fife, Scotland*, 2pp, no date, no author.

HORIZONTAL COMPOUND CORLISS ENGINE (1000 IND. HP.) FOR BOMBAY.
MR. GEORGE ALLAN, MEMB. INST. C.E., ENGINEER; MESSRS. DOUGLAS AND GRANT, KIRKCALDY, CONSTRUCTORS.
(*For Description, see Page* 414.)

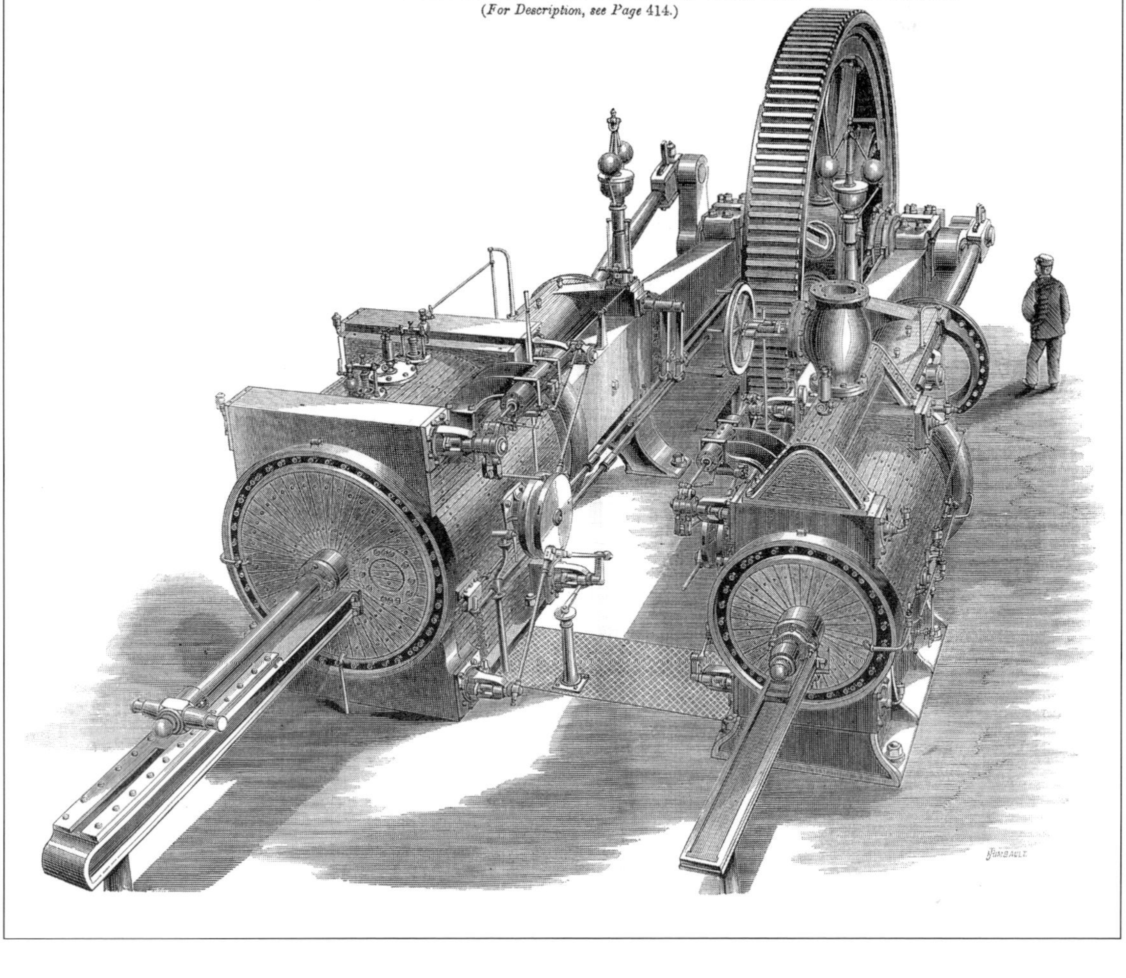

MAK/103 - Drysdale, Bon-Accord Works, Yoker, Glasgow.

Advertisements:
1909, 1910, 1956.

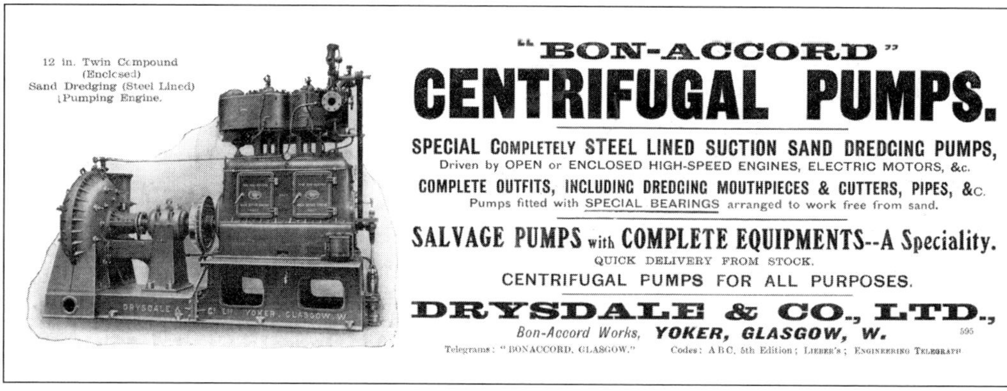

MAK/104 - Dujardin et Cie, Lille, France.

5) Printed material:
Article from *Engineering,* 31 August 1900 describing 1000-kW compound engine from the Paris Exhibition.

MAK/105 - D. J. Dunlop & Co., Port Glasgow.

5) Printed material:
Plate from *Engineering,* 7 January 1898, of vertical Corliss engine at Clonliffe flour mills (opposite page).

MAK/106 - Dunston Engine Works Co., Dunston-on-Tyne, England.

1) Advertisements.
1908.

MAK/107 - Easton & Anderson, Erith, Kent.

1) Advertisements:
Easton and Amos.
Copy of page from *The Artizan*, no page, no date, with illustrated advertisement for the patent hydraulic ram.

Easton, Amos and Sons.
Clipping from unidentified periodical, 22 February 1867, with ad for Patent Appold Centrifugal Pumps.

Copy of half page advertisement from unidentified periodical, no page, no date, with illustration for the Patent Appold centrifugal pump. This mentions several published reports on these pumps which GW does not seem to have been aware of.

VERTICAL CORLISS ENGINE AT THE CLONLIFFE FLOUR MILLS.

CONSTRUCTED BY MESSRS. DAVID J. DUNLOP AND CO., ENGINEERS, PORT GLASGOW.

(*For Description, see Page* 14.)

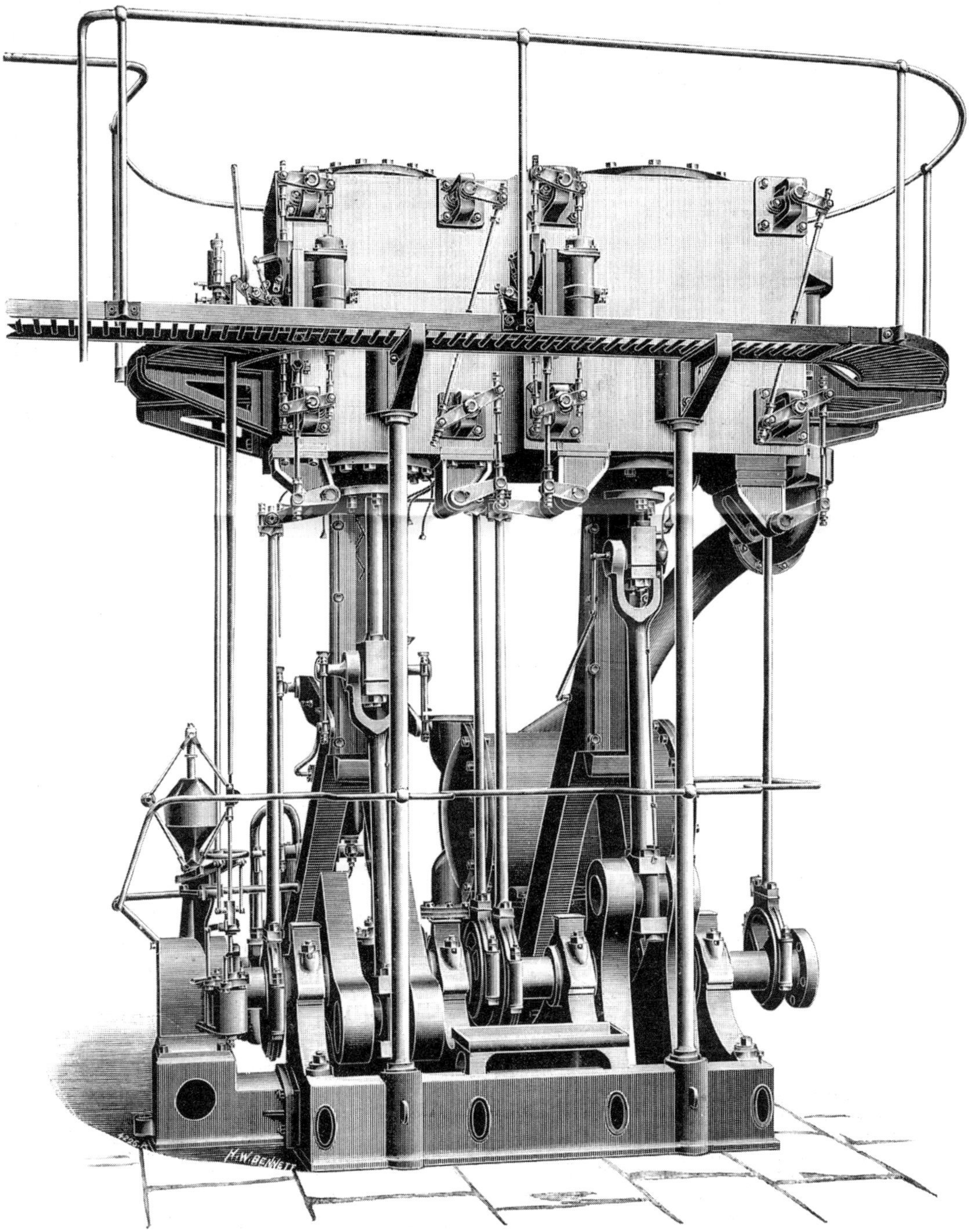

Easton & Anderson Ltd.
Full page from *Engineering,* 17 October 1890, with illustrated advertisement for beam pumps and gun mountings. [Some damage at side]. (see p. 108).

Reverse: 1890, full page advertisement for Greenwood & Batley, Ltd, Albion Works Leeds.

Easton, Anderson and Goolden.
Half page from *Engineering* 6 March 1896, with illustrated advertisement for triple expansion engine.

Reverse: 1896 Hudswell, Clarke & Co., Railway Foundry, Leeds. - shaft bearings.
Half page advertisement from unidentifed periodical, no page, no date, for patent electric coalcutter. With photocopy.

Reverse: n.d. Webster & Bennett, Machine Tool Makers, Coventry.

Photocopy of half page advertisement from *Engineering*, 1 January 1897, with illustrated advertisement for triple expansion engine. (with photocopy)

Same side: 1897 S. Owens & Co., Whitefriars Street, London EC, - improved compound duplex pumps.

Half-page advertisement from unidentified periodical, no page but 1897 for Erith high speed engine.

Reverse: 1897 Thomas Piggott, Birmingham. - welded and riveted pipes.

Half-page advertisement from unidentified periodical, no page but 1900 for Erith high speed engine and dynamo.

Reverse: 1900 Smith & Grace, Thrapston - pulleys etc.

Identical advertisement to above but 1901, with photocopy copy.

Photocopy of page from *Cassiers Magazine*, no page, no date with illustrated advertisement for dynamo.

Same side; n.d. Walter Dixon & Co., - multiphase motors.

Easton and Company.
Half page advertisement from unidentified periodical, no page but 1901 for electric lift gear.

Reverse: 1901 Smith & Grace, Thrapston, Pulleys etc.

Quarter page advertisement from unidentified periodical, no page, no date, about Schmidt's superheated steam engines.

2) Letters.
Sequence of letters between GW and Sigmund Pulsometer Pumps Ltd.

Carbon of letter from GW to the firm, dated 23 March 1976 asking about their records.

Letter to GW from the firm, dated 14 April 1976, acknowledging his letter of 23 March.

Letter to GW from the firm dated 9 March 1977 about his borrowing their file of the take-over of Eastons by Pulsometer.

Letter from GW to the firm in response to the above, September 1977 and offering to lend it to him.

Sequence of letters between GW and the Borough Librarian Southwark, about the siting of Easton's original works.

Letter to GW, dated 12 April 1978, with carbon of GW's reply on reverse, dated 10 June. Enclosed copies of various maps of the area.

Letter to GW, dated 4 May with note of a directory entry. Includes copy of map.

Sequence of letters between GW and Louis Kelting (Bridgwater) about Appold's patent.

Letter to GW, dated 17 August 1976 about Appold taking over an American patent [Lloyds of 1848].

Carbon of GW's reply citing his source for this.

Sequence of letters to GW from J. G. B. Hills, Production Editor of Newcomen Society.

Letter to GW dated 13 May 1979 with photocopy of text correction to Watkins's lecture in *TNS* 48. Includes a copy of advertisement for Green & Carter Ltd, Vulcan Ironworks, Kingsworthy, Hants, successors to Easton, Courtney & Darbyshire, who took over Easton's hydraulic ram business. The advertisement said they had all Easton's records of the ram business, which could well go back to the time of James Easton and the Montgolfier patent. Carbon of GW's reply on reverse, dated 17 May 1979, saying he had never heard of Easton, Courtney & Darbyshire.

Sequence of letters between GW and J. W Chase, 6 Gurney Close, Caversham, Reading.

Letter to GW, no date, offering him hospitality if he ever wanted to come to Reading to do work on the Pulsometer records.

Letter to GW, dated 4 November 1978, saying he had just bought from a local dealer a collection of Easton & Anderson's original drawings. They dated from 1888 - 1903, and comprised 69 drawings in all. Includes detailed list. Had paid £5 for them and was offering them to GW at the same price.

3) MS material:
Collection of MS and TS notes, drafts etc, for the short Newcomen Society lecture GW gave at Birmingham 14 May 1977.

Collection of MS and TS notes about the founders and company changes, notes on later business customers.

Includes the following printed biographies.

James Easton, from *The Engineer*, 3 November 1871.
Sir William Anderson, from *Proc. I.Mech.E.*, October 1898, pp 696 -701.
Charles Amos, from *Proc. I.Mech.E.*, January 1883.
Walter Goolden, from *The Electrical Review*, September, 1901. p 521.

There is also a collection of copies of original material from the records of the Pulsometer Co. at Reading, all dating from 1908 when the firm finally was taken over. Includes detailed financial summaries of profits and costs etc.

There is a also a set of photocopies of pages from the Easton and Anderson engine records at Reading, by courtesy of Mr Chase, qv, in letters above.

4) Photographs:
Set of 4 prints of the Oakey beam engine (SER 328), with stamp of Larkin Bros, 210 High Holborn, London WC1, neg. no. A 58538 - 58541.

5) Printed material:
Sequence of copies from printed accounts of Easton's products, from the technical press, and technical books.

Easton and Amos.
Article from *Civil Engineer & Architects' Journal,* 1849, pp 165-66, about a Simpson WW pumping engine, with improved pump.

Letter from *Civil Engineer & Architects' Journal,* 1849, pp 213-214, with letter from Easton & Amos about the pump in 1) above.

Extract from Jabez Hogg, surgeon, *Elements of Experimental and Natural Philosophy.* London 1853, pp 140-143, with detailed account of the Appold Pump at the Great Exhibition.

Copy of Amos's patent, No 2802 of 1857. (Lacks drawing).

Extract from *The Engineer*, 11 June 1858, (page no. cropped), with illustrated account of Amos's patent engine. [2 copies].

Extract from *The Engineer*, 19 May , 1862, p 302, with illustrated account of Easton, Amos, and Sons' Appold draining machine. [2 copies].

Extract from *The Engineer* 19 June 1863, p 346, with illustrated account of Easton and Amos's machinery for sawing logs.

Extract from *The Engineer*, 28 August 1863, p 125, with illustrated account of Amos's machinery for the manufacture of paper.

Extract from *The Engineer*, 20 April 1866, p 279, with illustrated account of centrifugal pumps and plans.

Easton, Amos & Anderson.
Extract from *The Engineer*, 26 October 1866, pp 312-313, with illustrated account of double pair of oscillating engines for the Emperor of Abyssinia. [with 4 copies of the drawing].

Extract from *The Engineer,* 9 November 1886, pp 363, with illustrated account of steam boilers for the Emperor of Abyssinia. [2 copies).

Extract from *The Engineer*, 16 November 1866, p 381, with illustrated account of floating centrifugal drainage pump.

Extract from *The Engineer*, 21 December 1866, p 488, with illustrated account of 30-ton double-action travelling cranes.

Extract from *The Engineer*, 9 May 1879, p 336, 339-40, with illustrated account of gun carriages for the Russian *Cyclads.*

Extract from *The Engineer*, 21 August 1891, p 206, with un-illustrated account of Anderson's revolving purifier for river water.

Extract from *The Engineer*, 8 January 1892, pp 42-3, with illustrated account of high-speed engine and dynamo. [2 copies].

Extract from *The Engineer*, 6 May 1892, p 383, with illustrated account of electro-motive car, Bradford Tramways.

Easton, Anderson & Goolden.
Extract from *The Engineer*, 15 February 1895, p 144, with illustrated account of the engines and centrifugal pumps, Victoria Docks, West India Dock, London.

Extract from *The Engineer*, 10 January 1896, pp 33-34, with illustrated account of pumping engines, Colchester Waterworks.

Extract from *Engineering*, 1 January 1897, p 25, with illustrated account of feed, check valves and blow-off cocks.

Extract from *Engineering*, 12 March 1897, p 340, with illustrated account of Niger stern wheelers *Empire* and *Liberty*.

Extract from *The Engineer*, 11 June 1897, p 599, with illustrated account of E. A. & G.'s exhibit at the Electrical and Engineering Exhibition, Newcastle upon Tyne. [Not complete].

Extract from *The Engineer*, 21 October 1898, p 393?, with illustrated account of hydraulic pumping engines for Tasmania [2 copies].

Extract from Frederick Colyer, *Pumps and Pumping Machinery* 2nd ed, 1889, pp 14-17, with plate, of Easton & Anderson duplex pump.

Extract from *The Engineer*, 6 January 1899, with illustrated account of the electric lifts for the New Brighton Eiffel Tower.

Extract from *The Marine Engineer*, 1 October 1899, pp 307-310 with illustrated account of Easton, Anderson & Goolden's Erith Iron Works.

Extract from W. Norris and Benjamin H. Morgan, *High Speed Engines*, 1900, pp 37-40, with illustrated account of high speed engines made to Bjornstad's design.

Easton & Co.,
Extract from *The Engineer*, 30 January 1903, p 125, with illustrated account of 300 HP tandem compound engine for superheated steam. (Schmidt patent). 12 copies].

Extract from *Engineering*, 27 May 1904, pp 735-736, with illustrated account of the Antwerp Water Works. [lacks some text], NB There is evidently a continuation to this which is lacking from the collection. Mentions Easton, Courtney & Darbyshire, of Westminster, consulting engineers.

Extract from *Engineering*, 10 June 1904, p 809, with illustration of the Antwerp Water Works; Pumping Station at Luythaegen. [Lacks descriptive text].

Easton and Anderson branch of the Pulsometer Co., Reading.

Extract from *The Engineer*, 30 January, 1915, (page unknown) with illustration of the steam pumps at the West Gloucester Water Co., [all on small sheets with perished Sellotape, Lacks any descriptive text].

Extract from *Model Engineer & Practical Electrician*, Vol 73, 14 November 1935, pp 463-464, with illustrated account of the final run of the Ashridge beam engine. The engine had been out of use several years previously, but on learning that it was to be scrapped a team of the Aylesbury Gang, aided by enthusiasts from all over the country, arranged for a farewell steaming. GW in fact visited Ashridge in 1955 and photographed the engine.

6) Trade literature:
Photocopy copy of Easton, Anderson & Goolden's catalogue for 1895, ex the Pulsometer Co., (See letters about this, above).

Note: Lacks the following pages: 5, 22, 40-43, 57, 78, 79, 83, 92, 93, 96, 97, 99.103, 111, 114, 115 (duplicates of 3, 68).

Photocopy of the same catalogue, but Science Museum's copy, presented April 1951. 2 original pages per page. Note: Lacks the following pages: 56, 57, 66, 67.

List of the Easton and Company's customers, 1908. [includes many from the earlier incarnations of the firm, and repeats many of the names listed in the catalogue above.

7) Other material:

Collection of 69 original works drawings of Easton and Anderson's 1888-1903 (see Section 38, Large drawings and photographs, for more details).

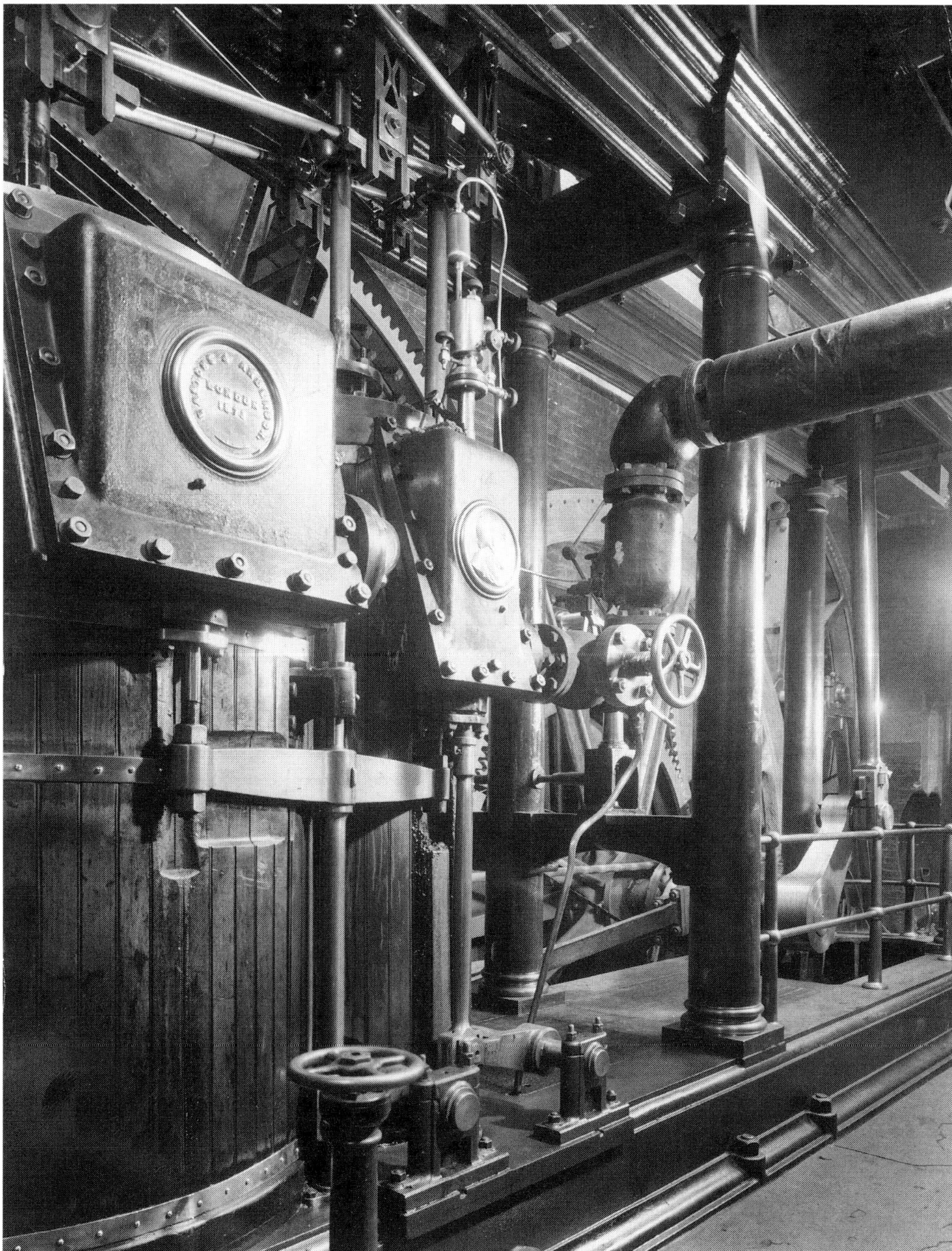

Above & opposite page: The Oakley Beam Engine, (see p.103).

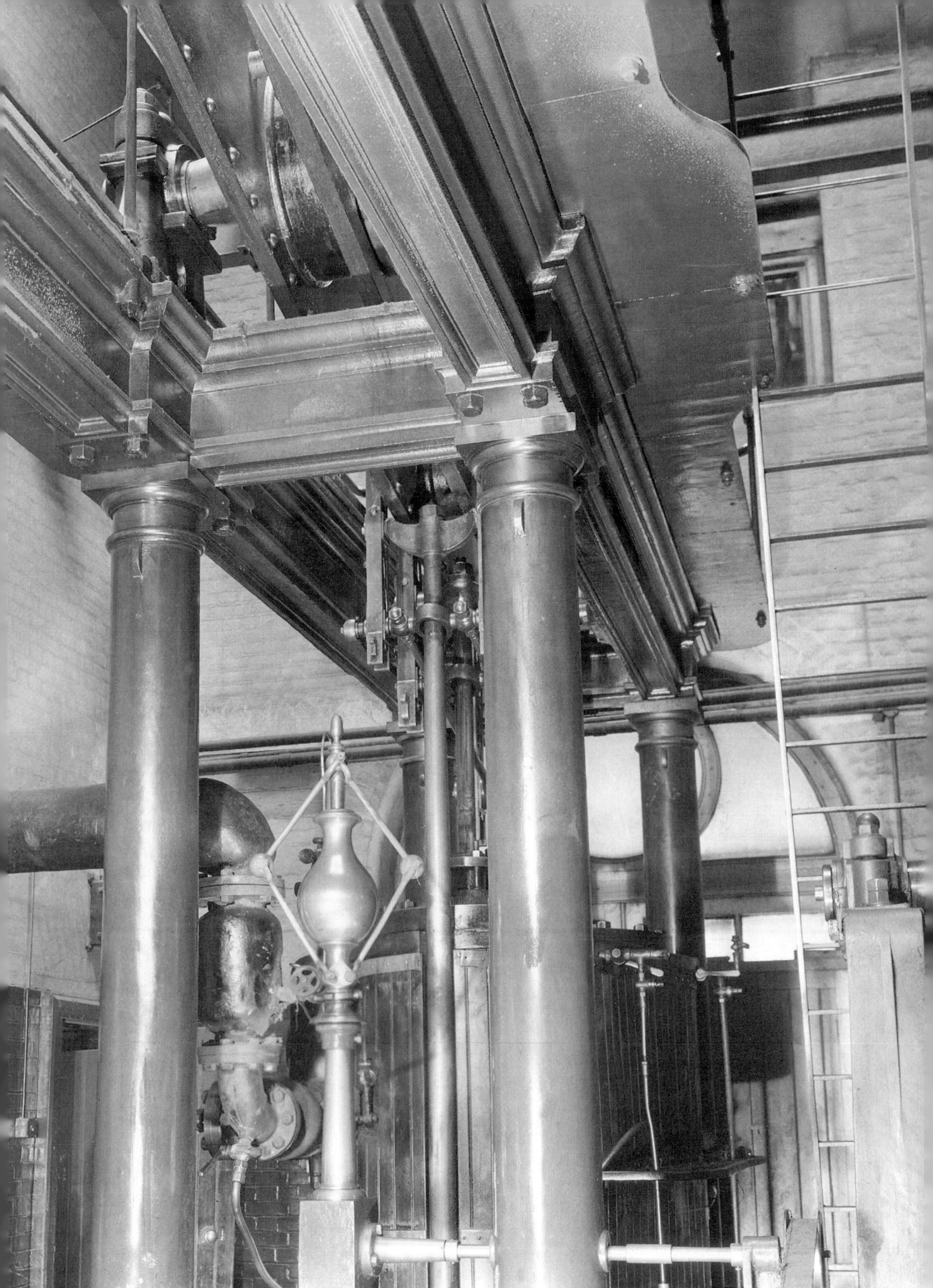

Oct. 17, 1890.] ENGINEERING. [SUPPLEMENT page XV] 41

EASTON & ANDERSON, LTD.

LONDON & ERITH.

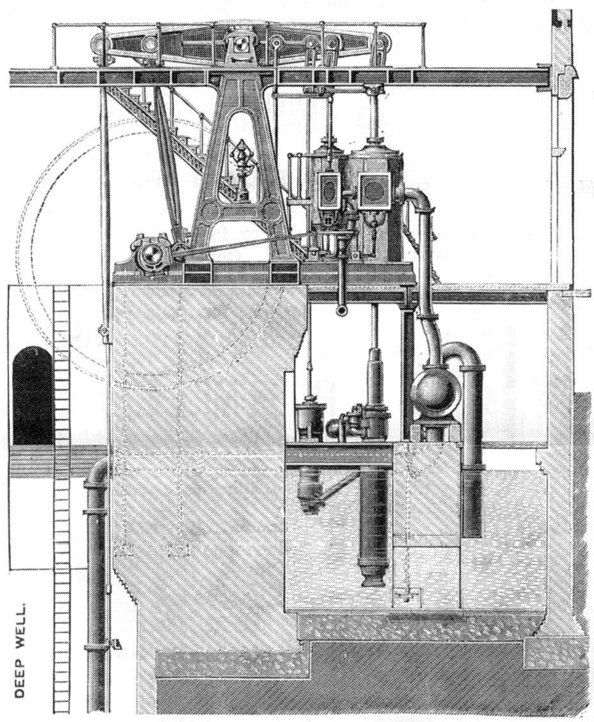

BEAM ENGINES—VARIOUS SIZES.

MANUFACTURERS OF

PUMPING, DRAINAGE, & MILL ENGINES.

HIGH-SPEED ENGINES

FOR

Dynamos, Centrifugal Pumps, and Ventilating Fans.

COMBINED ENGINES & CENTRIFUGAL PUMPS.

BOILERS.

HYDRO-PNEUMATIC & OTHER GUN CARRIAGES.

HYDRAULIC LIFTS AND PRESSES.

HYDRAULIC AND STEAM CRANES.

WATER WORKS, &c.

FIRE SERVICE & HEATING APPARATUS

FOR ESTATES AND PUBLIC BUILDINGS.

ANDERSON'S PATENT REVOLVING WATER PURIFIER.

LEEDS PATENT FILTER.

PAPER AND SUGAR MACHINERY.

CAISSONS AND DOCK GATES, &c., &c., &c.

OFFICES:

3, WHITEHALL PLACE, S.W.

Telegraphic Address: "Egyptian, London." Telephone No. 3695.

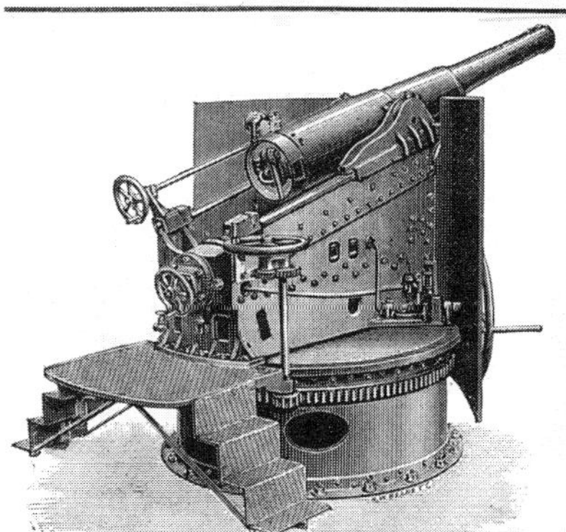

ANDERSON'S PATENT STORAGE RECOIL MOUNTINGS.

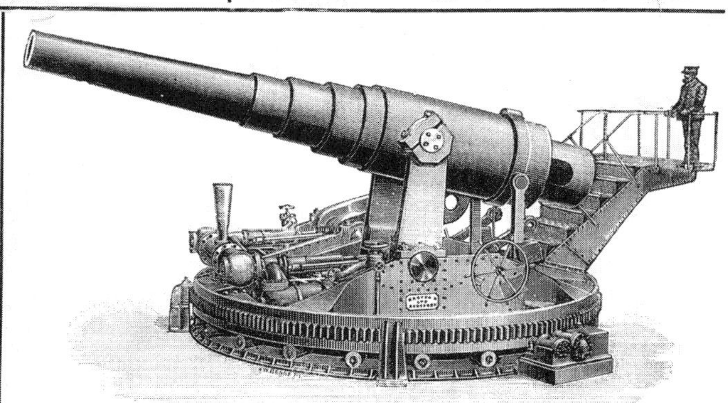

DISAPPEARING MOUNTINGS FOR TWO 50½ TON BREECH-LOADING GUNS.

8589

MAK/108 - Easton & Bessemer, Taunton, Somerset.

1) Advertisements:
1903, 1904, 1905 (3), 1907 (2), 1908.
Reverse: 1903 George Scott & Son, London - air compressors.
Reverse: 1905 S. Howes - *Little Giant* water turbine.
Reverse: 1908 Measures Bros, London & Croydon - constructional steelwork.

5) Printed material:
Article from *The Engineer* dated 19 February 1904 describing a horizontal compound engine 12 in and 22in x 28in. (see below).

COMPOUND HORIZONTAL ENGINE
EASTON AND BESSEMER, LIMITED, ENGINEERS, TAUNTON

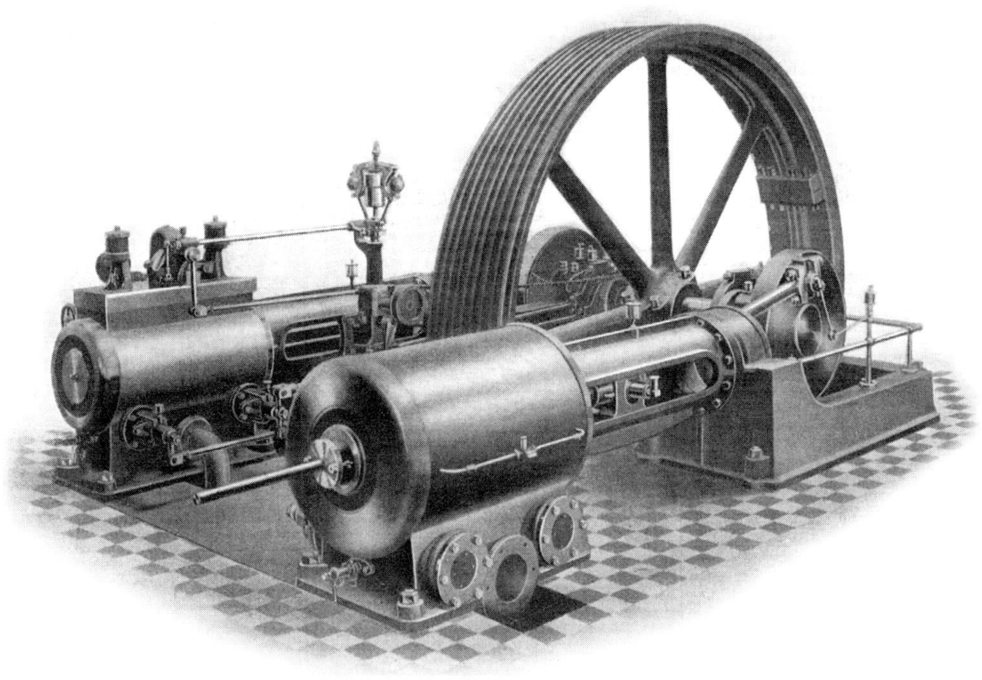

MAK/109 - Easton & Johnson, Taunton, Somerset.

6) Trade literature:
Photocopy of 18 pages from undated catalogue: c 1920.

MAK/110 - J. Edge & Co., Tipping Ironworks, Bolton.

5) Printed material:
Faded photocopy of illustrated leaflet of horizontal engine 1870.
Faded photocopy of part of article from *Engineer* 18 July 1873 p 37. lacks plate.
Faded photocopy of p 272 of *Science of Modern Cotton Spinning*, no date, no author, illustrating Edge's patent steam engine.

MAK/111 - Ehrhardt & Schmer, Saarbrücken, Germany.

5) Printed material:
Photocopy of article from undated page from *Iron & Coal Trades Review*, 18 Oct 1907 about 20,000 hp 3-cylinder reversing rolling mill engine.

MAK/112 - Elliott, Pittsburg, USA.

6) Trade literature:
Pages 9-10 from catalogue of Unaflow engines (Stumpf patent). No date.

MAK/113 - Erie City Iron Works, Erie, USA.

6) Trade literature:
Photocopy of pages from catalogue of steam engine boilers and saw mill machinery of 1879, 11 pages.

Booklet illustrating engines, boilers, feedwater heaters and tanks. 1916, 40 pp.

MAK/114 - Escher, Wyss & Co., Zurich.

2) Letters:
Letter from Arthur Stowers to GW about photocopy of pages from Escher Wyss book (see 7 below).

5) Printed material:
Article from *Engineering*, 7 September 1900 about 1000 hp horizontal engine at the Paris Exhibition.

Article from *Engineering*, 28 January 1901 about 300 hp triple-expansion engine.

Article from ?*Mechanical World*, [top cropped so no date or page], about compound engine with Frikart valve gear at the Paris Exhibition of ? 1901.

7) Other material:
6 pages copied from *Escher, Wyss, 1805-1955, 150 years of development.* (See Stowers' letter, noted above).

MAK/115 - J. Evans, Cullwell Works, Wolverhampton.

1) Advertisements:
1903 (2), 1905 (2).

6) Trade literature:
Book - *Section 2 Pumps and Pumping Machinery* 1894, 3ed. rev to 1904, 160 pp.

MAK/116 - Fairbairn & Co., Manchester.

2) Letters:
4 pages of MS notes about Sir William Fairbairn and Fairbairns.

5) Printed material:
Obituary of Sir William Fairbairn from *Proc. I.Mech.E.*, 1875, pp 22-24.

Illustrated account of Fairbairn's works from *The Commercial Aspect of Manchester*, no author but perhaps Measham, n.d. but c 1850.

7) Other material:
Photocopy of plate of Fairbairn side-lever engine from Scott Russell's *Treatise on the Steam Engine*, c 1850.

MAK/117 - *Fairbanks, Morse & Co., 126 Southwark Street, London, S. E.*

1) Advertisements:
1905 for compound duplex outside centre pumps.

MAK/118 - *Joseph Farcot, St Ouen, France.*

5) Printed material:
Extract from *Mechanical World*, 15 March 1890, pp 106-107 describing a horizontal Corliss engine at the late Paris Exhibition (Part 1 only).

MAK/119 - *Fawcett, Preston & Co., Engineers, Liverpool.*

1) Advertisements:
1908, 1911.

5) Printed material:
Article from *Dorset County Magazine* No 57, 1976 describing rescue of Fawcett, Preston horizontal twin at brickworks of Chas. Mitchell & Co., Poole, Dorset.

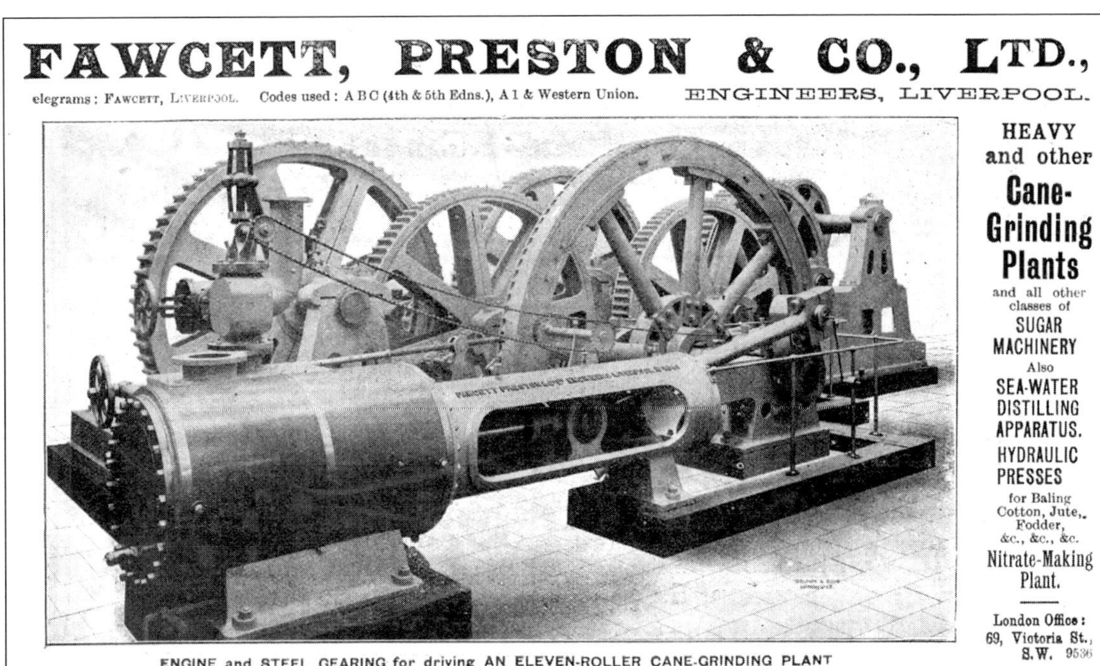

MAK/120 - S. Z. de Ferranti Ltd., Hollinwood, Lancs.

4) Photographs:
2 prints of 3000 hp engines at Deptford, 1906 (below & opposite: latter endorsed 'Monorell [?] governor fitted to deal with the LBSC Ry Traction load').

Print of Ferranti engine at Dover, 1936.

Print of 1000 kW South London Electric Supply Co.'s engine, pencil note on back Z224, n.d. , see below ?Copied from this article.

5) Printed material:
Extract from *Engineering* 19 October 1900, pp 509-512 with article about 1000 kW engine and alternator at South London Electric Supply Co.'s station near Loughborough Junction [between Brixton and Camberwell].

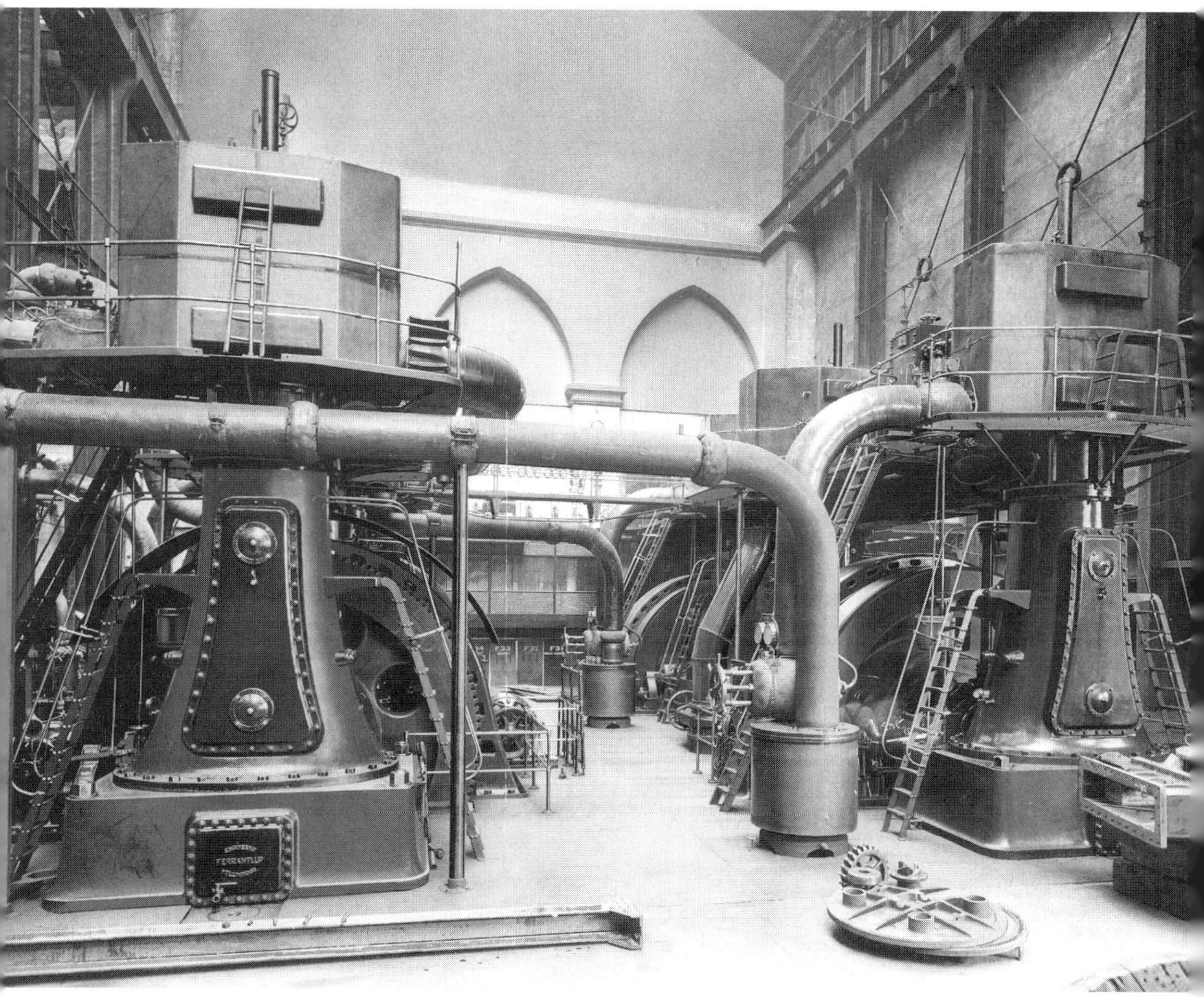

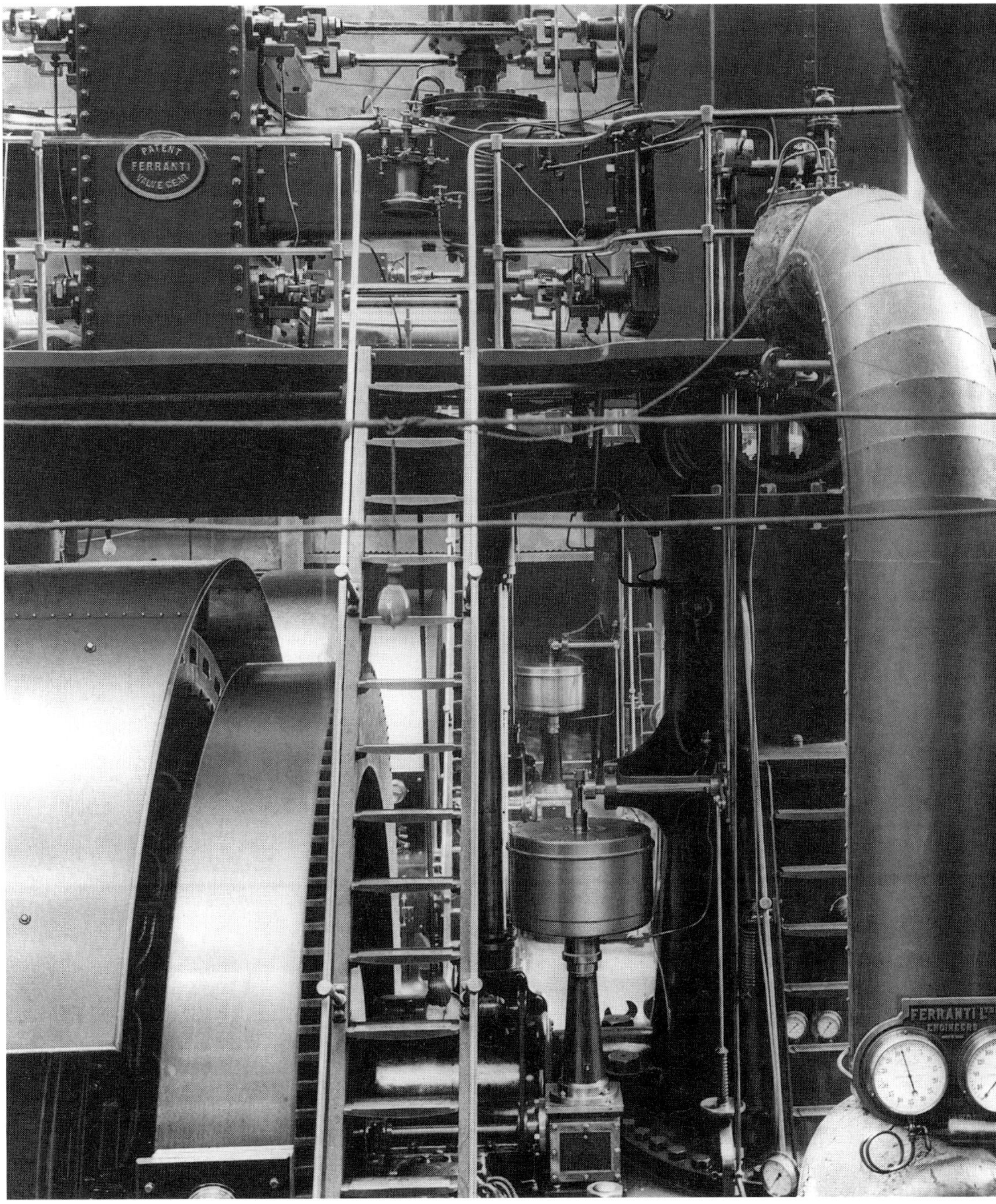
PATENT
FERRANTI
VALVE GEAR
FERRANTI Ltd
ENGINEERS

MAK/121 - Filer & Stowell, Milwaukee, Wis. USA.

6) Trade literature:
26 page illustrated booklet describing the products of the firm, n.d. but c 1940.

MAK/122 - Fleming & Ferguson, Paisley.

1) Advertisements:
1894, 1896 (2), 1897 [One of the 1896 ads is full page].

***2) Letters*:**
Letter to GW from Fleming & Ferguson dated 22 July 1953 in response to a query about their triangular connecting-rod engines.

Letter to GW from Fleming & Ferguson dated 19 June 1963 saying they had no records available.

Letter to GW from F. G. Bettles of Taunton, dated 3 August 1963 regarding a model he had made of a Musgrave triangular connnecting-rod engine with 2 prints of the engine and a photocopy of the Musgrave engine.

Letter to GW from Bill Lind dated 20 September 1972 referring to the photocopy copies noted (*) below and asking GW's assistance in tracing one of the engines mentioned.

5) Printed material:
Article from *Engineering* 1 January 1897 describing a single crank quadruple expansion mill engine and boiler. (see p.115-116).

* Article from *The Engineer* 19 April 1889, describing the quadruple expansion engines of SS *Singapore* [drawings only] - Triangular connnecting-rod, non-dead centre type.

* Photocopy of letters from *The Engineer* 10 May 1889 about the engines of SS *Singapore* in response to previous article, claiming earlier priority for the invention of this design.

* Page from *Shipbuilding and Shipping Record*, 7 Aug 1941 with letters about the Fleming & Ferguson marine engines then displayed in the Smithsonian Institution. See Bill Lind's letter above.

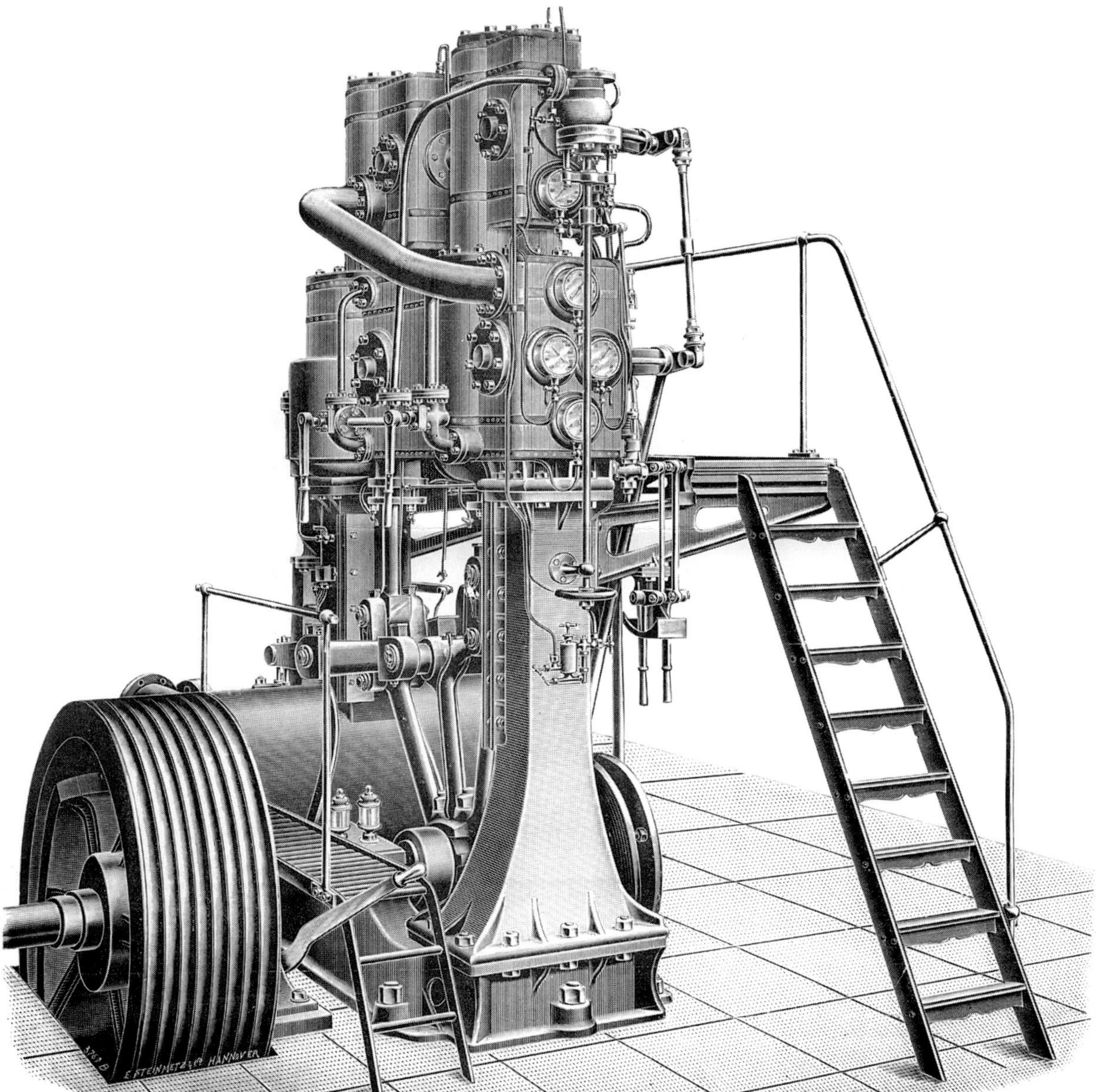

Above & p.115: Illustrations of article in *Engineering*, 1 Jan 1897 describing a single crank quadruple expansion mill engine and boiler by Fleming & Ferguson.

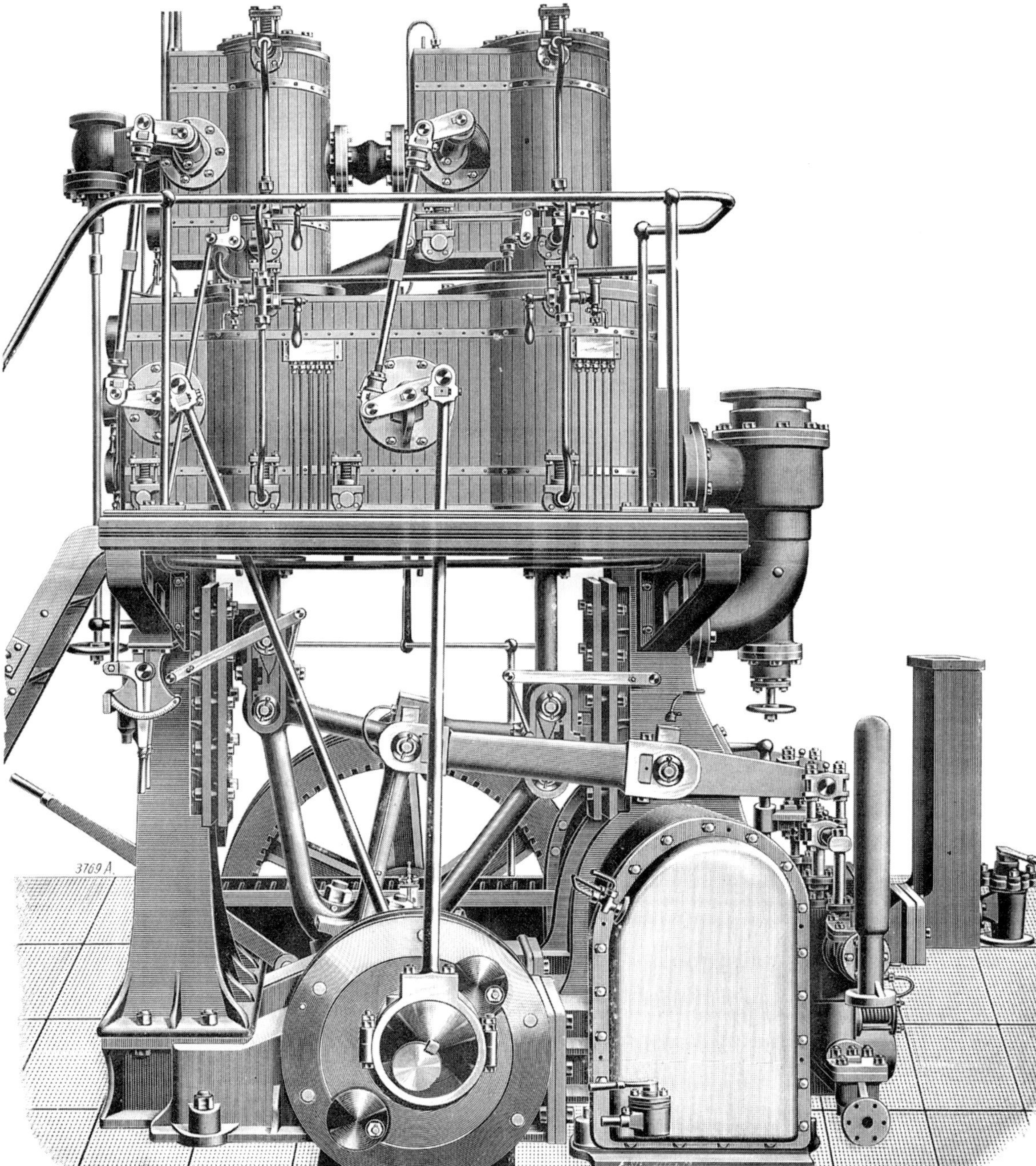
3769 A.

MAK/123 - E. Foden, Elworth Foundry, Sandbach, Cheshire.

2) Letters:
Letters to GW from C. K. Hockenhull [of Foden's drawing office] dated 31 May and 27 June 1956, 23 May 1968, 2 February 1959, relating to photos and drawings of their stationary engines, esp. South Wales rolling mill engines.

3) MS material:
MS abstract of details of stationary engines from an early cost ledger, 1875 onwards, 3 pp [in handwriting of Mr Hockenhull).

4) Photographs:
Print of rolling mill engine at the South Wales Tinplate Co., Port Talbot, running post - 1946. (see below).

7) Other material:
Photographic copies of GA drawings of:

28in and 52in inverted twin compound condensing engines 3ft 6in stroke for Clayton Tinplate Co., South Wales (n.d.).

24in and 46in horizontal compound condensing engine 5ft stroke (n.d.) Coupled 16in horizontal winding engines (9 August 1872).

26in hauling engine, 4ft stroke for Ashton Moss Colliery, Audenshaw (Apr 1891).

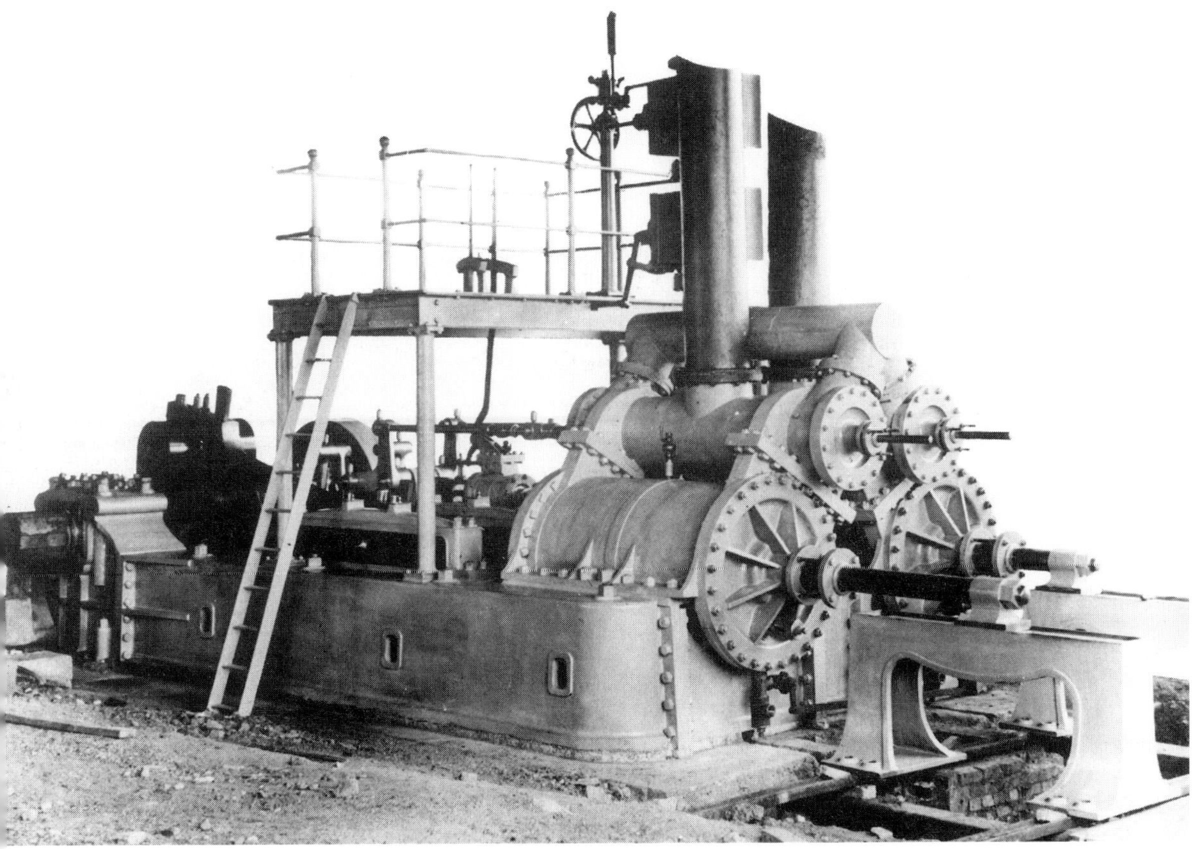

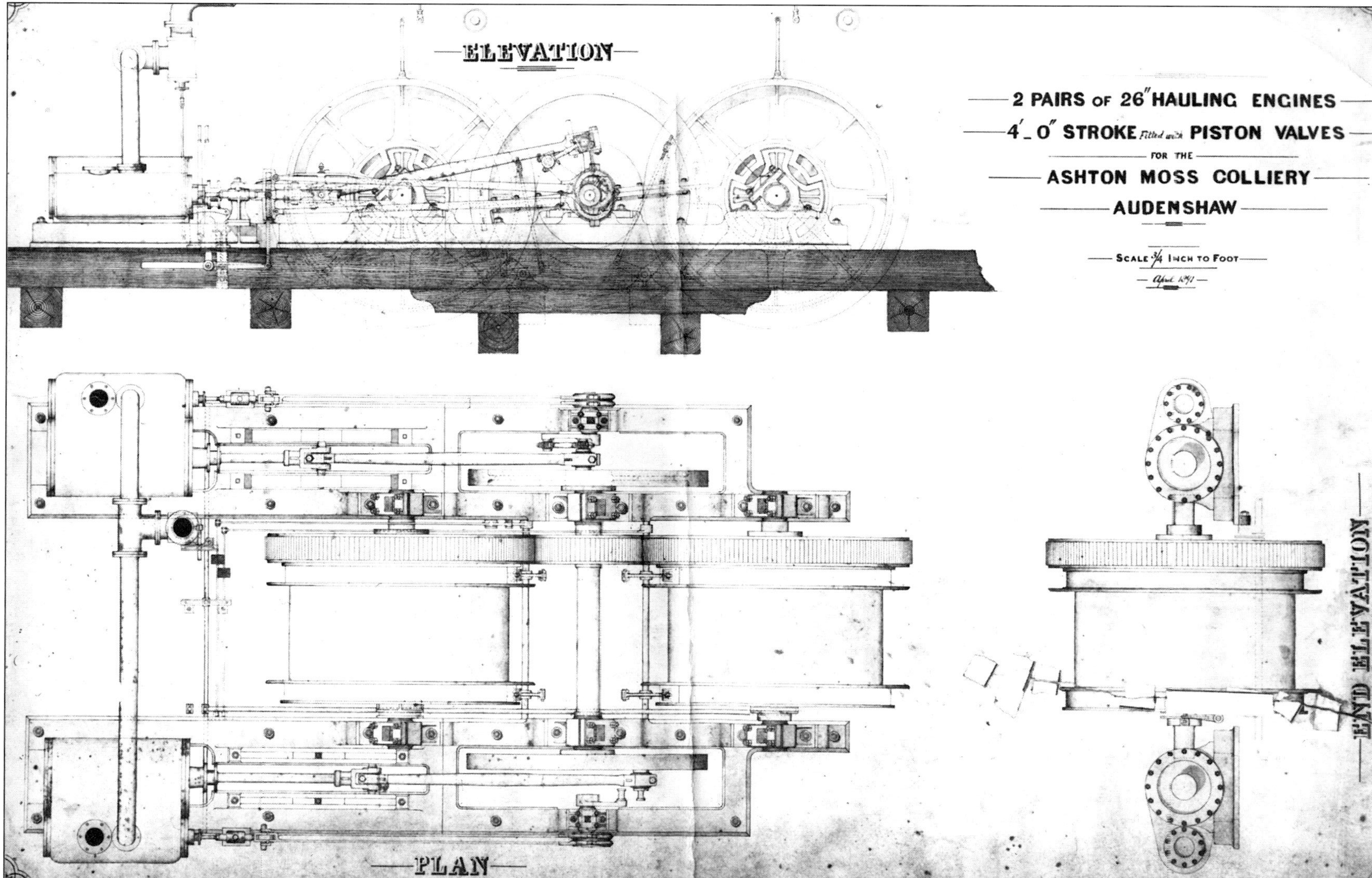
ELEVATION
2 PAIRS OF 26" HAULING ENGINES
4'_0" STROKE Fitted with PISTON VALVES
FOR THE
ASHTON MOSS COLLIERY
AUDENSHAW
SCALE 3/4 INCH TO FOOT
April 1891
PLAN
END ELEVATION

MAK/124 - *Joseph Foster, Preston, Lancs.*

5) Printed material:
Article from *Mechanical World*, dated 8 January 1892, pp 18-19 describing a triple expansion beam engine for Kent Street Cotton Mills, Preston, see below.

Photocopy of page from catalogue n.d., describing Foster, Fails & Thorn 1250 hp uniflow for mill in ?Ahmenabad, India, 1934.

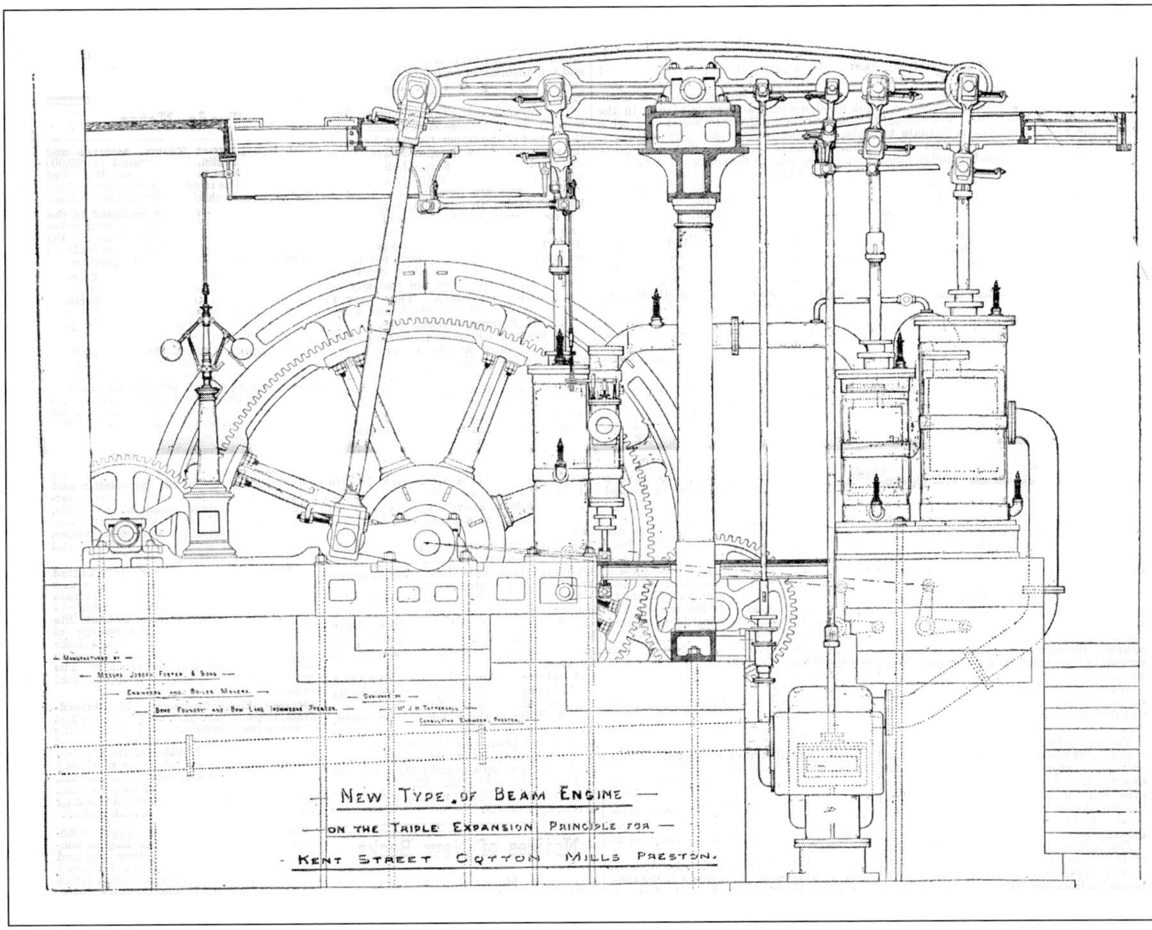

MAK/125 - William Foster & Co., Ltd, Wellington Foundry, Lincoln.

1) Advertisements:
1894 (see below).

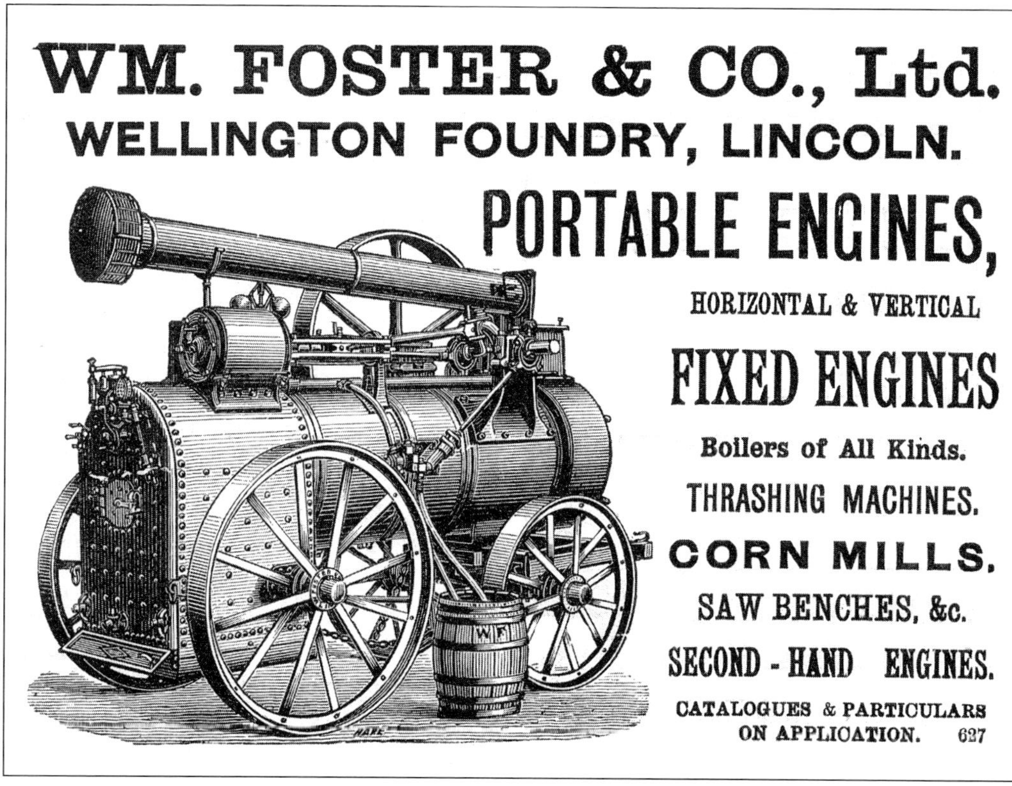

MAK/126 - Fowell & Son, St Ives, Hunts.

7) Other material:
Preprint of Newcomen Society lecture 'The Fowells of St. Ives', by T. B. Paisley, 1964.

MAK/127 - John Fowler & Co., Leeds.

1) Advertisements:
1887, 1888, 1892, 1901, 1911.

Reverse 1887 W. M. Melville, Sinclairtown Foundry, Kirkcaldy; Richard Garrett, Leiston.
Reverse: 1888 Bell's Asbestos Packing, 118, Southwark Street, London SE.
Reverse: 1892 Stothert & Pitt, Bath; John Musgrave & Sons, Bolton.
Reverse: 1901 Galloway's, Manchester.
Reverse: 1911 W. H. Bailey & Co., Salford, Manchester - recorder makers.

2) Letters:
Letter to GW from Fowlers dated 21 February 1950 about their catalogues.

Letter to GW from Robert Polehampton dated 16 June 1980 - not relevant to Fowlers (see main Polehampton letter file).

4) Photographs:
Print of ?Fowler portable at Bodleys, Exeter, n.d.

5) Printed material:
Pages 342-342 from Haeder & Powles on Fowler engines.

Pages from *Model Engineer* 4 May 1979 'Jeynes Corner' about the Newcomen engine at Miora furnace - clearly misfiled.

Steaming Vol 23, No 2, March 1980, with lecture about Fowlers by Michael Lane.

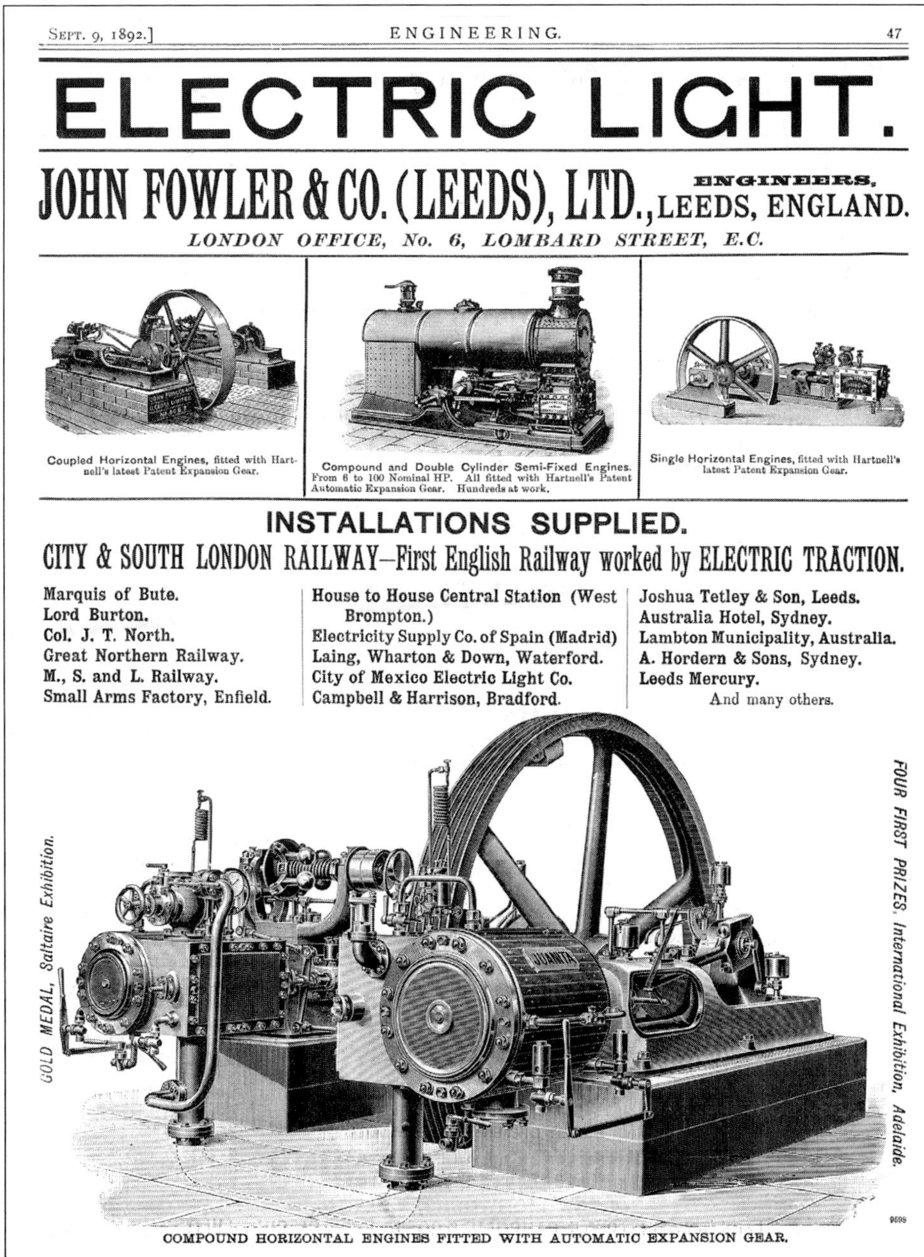

MAK/128 - Fraser & Chalmers, Erith, Kent.

1) Advertisements:

1895, 1896, 1897.

Reverse: 1895 George Richards & Co., Ltd, woodworking machine - tool manufacturers; Luke & Spencer Ltd, Broadheath, Manchester, - grinding and polishing machinery makers.

Reverse 1897 Cox & Co., Falmouth, - engineers & shipbuilders.

2) Letters:

Letter to GW from Michael Satow, OBE, dated 13 February 1979 with reminiscences of Fraser & Chalmers in the 1930s.

5) Printed material:

Photocopy of photograph from ?*Engineer* 14 Nov 1902 of air-compressor.

6) Trade literature:

Booklet: *General Catalogue:*, 1904, 96pp. (see opposite page).

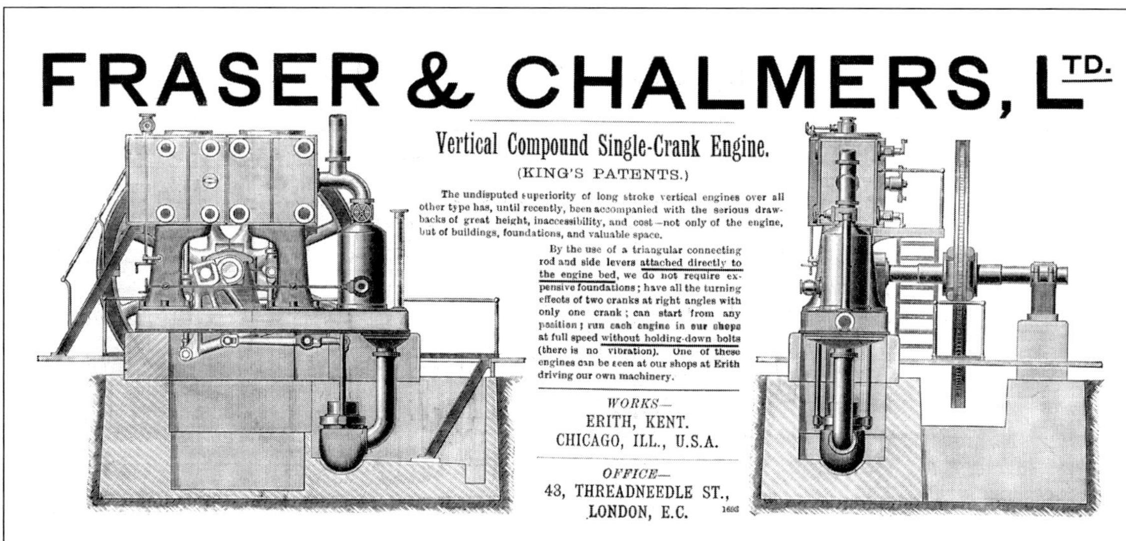

The 1896 Advertisement.

HORIZONTAL SLIDE-VALVE ENGINES

Plate 63.

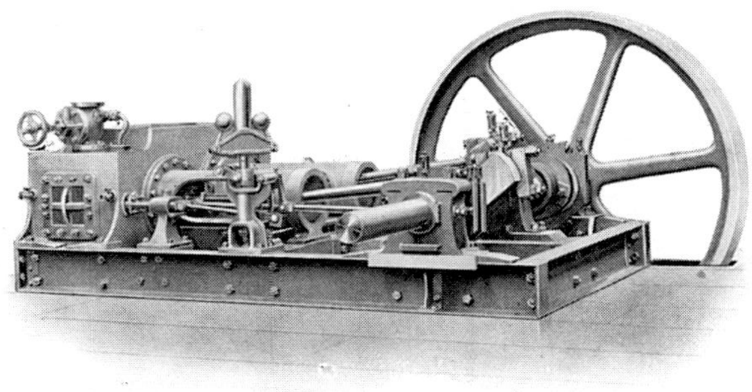

For smaller plants we supply Slide-Valve Engines fitted with Meyer cut-off gear, which have met with great approval.

Plate No. 63 shows a Cross-Compound Engine of portable self-contained type arranged on a girder frame.

Our Horizontal **Universal** Slide-Valve Engine has been specially designed for Mining and Colliery work. It is so designed that the engine can be erected to either hand. It is constructed with single, duplex, cross-compound, and tandem-compound cylinders, the single-cylinder engines being of sufficient strength so that they can be altered to any of the above combinations later. Where, as is often the case, an engine is required to be used for different purposes at different times with a possible increase in power, **a better engine than this cannot be obtained.**

Plate 64.

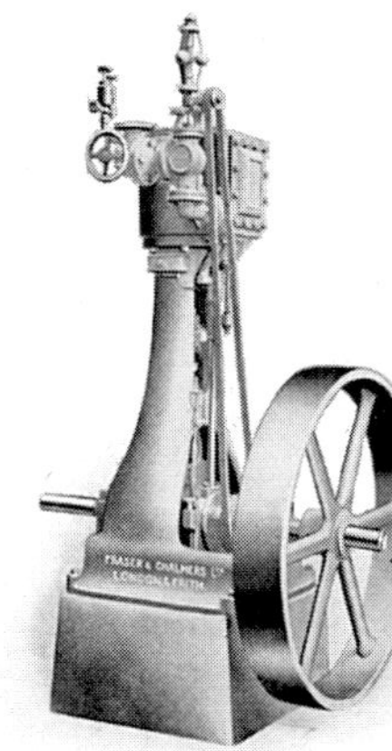

VERTICAL SLIDE-VALVE ENGINES

We supply a large number of small engines of this type.

For plants, such as cyanide plants, &c., where little power is required, they are exceedingly popular.

The small electric light equipments supplied by us usually have engines of this type direct connected to the dynamo.

Plate 65.

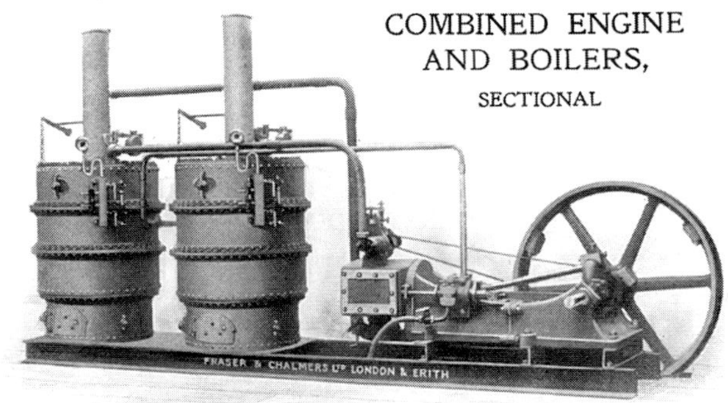

COMBINED ENGINE AND BOILERS,

SECTIONAL

We have had a long and varied experience in the design of sectionalized machinery for mining and prospecting work.

The above outfit is specially designed for easy erection and portability. On page 73 is shown a large pump sectionalized for difficult transport.

MAK/129 - Frick Co., Waynesboro, Pennsylvania, USA.

6) Trade literature:
A History of the Frick Company, Waynesboro, Pennsylvania, 1952, 46pp.

MAK/130 - Fryer & Co., Engineers, Rouen, France.

5) Printed material:
Article from *Engineering* 28 August 1903 describing the Lefevre four-cylinder high-speed engine.

MAK/131 - Fullerton, Hodgart & Barclay, Vulcan Foundry, Paisley.

1) Advertisements:
1905, 1911, 1912, 1913.

Reverse: 1911 John Lang, Johnstone, nr Glasgow - lathe manufacturer.
Reverse: 1912 H. W. Ward, 89 Lionel Street, Birmingham - lathe manufacturer.

2) Letters:
Letter to GW from Bill Lind dated 7 December 1977 describing the saving of the company records - included 1800 17in x10in glass plate negatives and 1500 film 10in x 8in. Sent GW some photocopies from early catalogues.

3) MS material:
Photocopied pages from order book dated 1846.

4) Photographs:
Envelope with 2 prints of engines at J. P. Coats' Mills, Paisley - one by Musgraves (see p.125), the other by Fullerton, Hodgart & Barclay (see pp. 127-128).

Envelope with collection of 12 works photographs of engines under construction, one each dated 1906, 1935, 1936; 3 prints from a catalogue: one dated 1899; small copy of print of the Fullerton, Hodgart & Barclay engine at Coats' Mill, Paisley, noted above.

6) Trade Literature:
Photocopied pages from an early catalogue:, n.d. but c 1910, interleaved with contemporary MS notes about the engines illustrated.

Catalogue of Steam Engines and Condensers, 22pp, n.d. but c 1914.

Catalogue of Hydraulic Plant, 42pp, n.d. but c 1914.

7) Other material:
Print of ?Fairbairn's works - ?misfiled.

MS list of engines built by the Ebor Engineering Co., Littleborough ?misfiled.

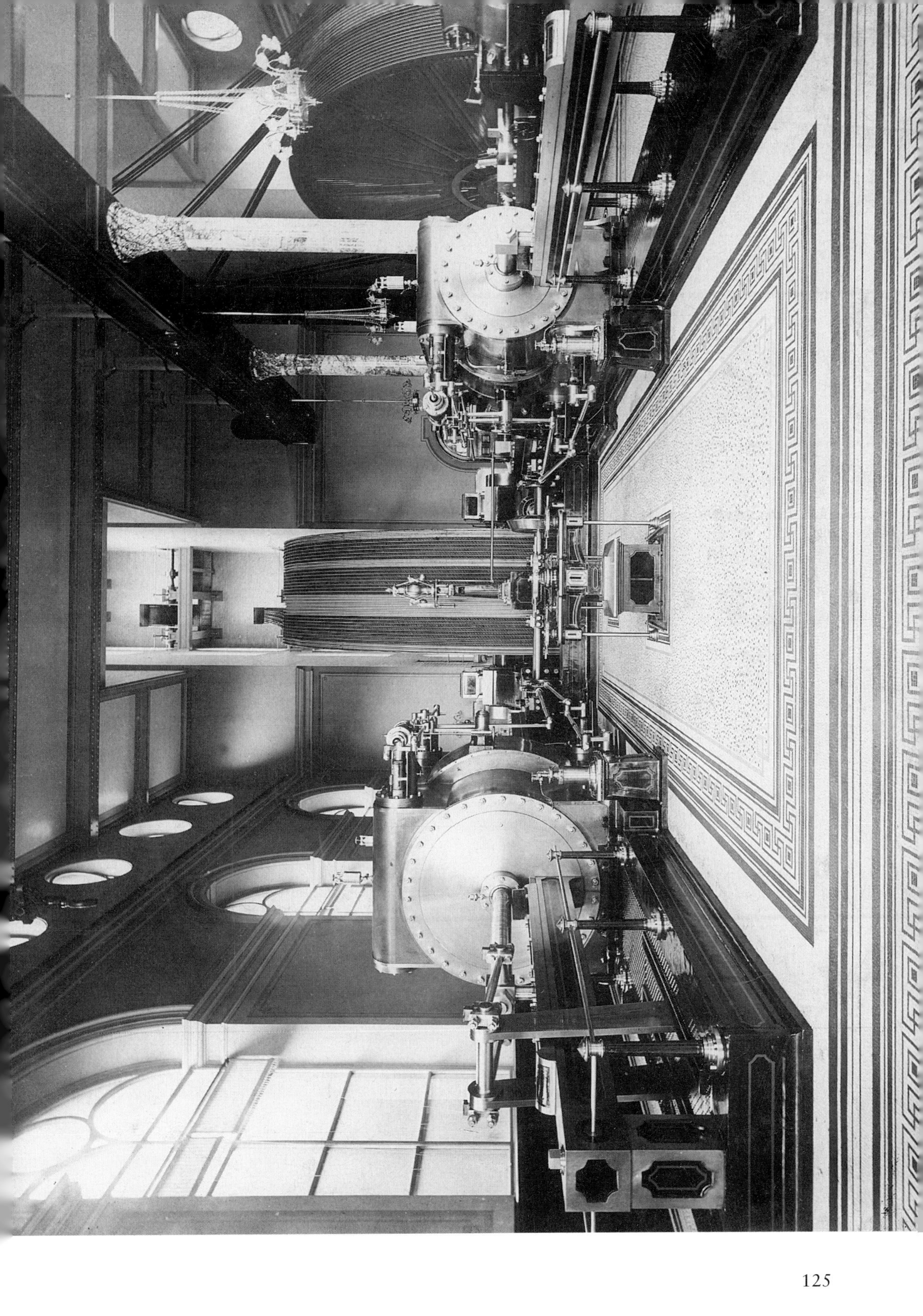

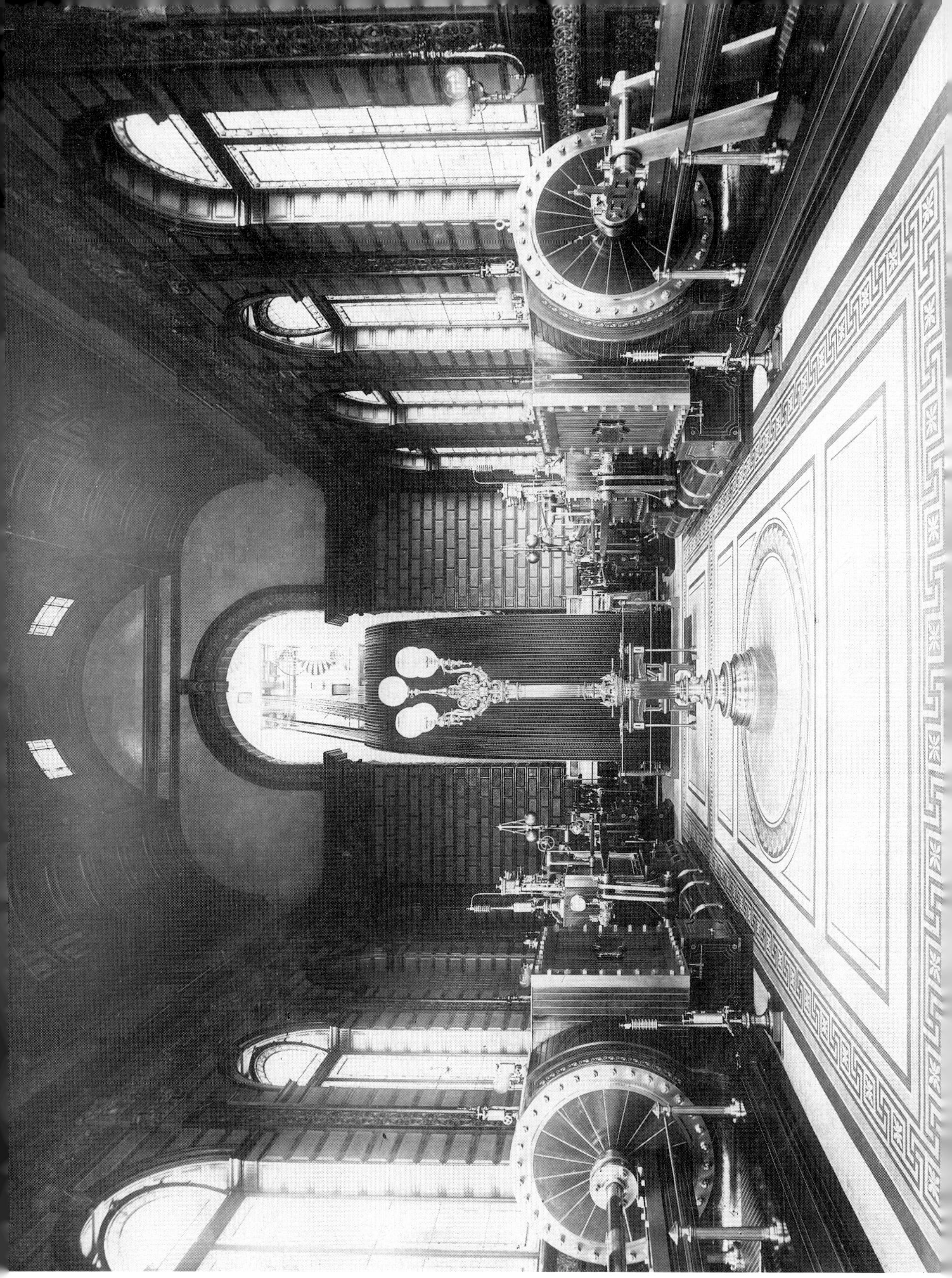

Fullerton

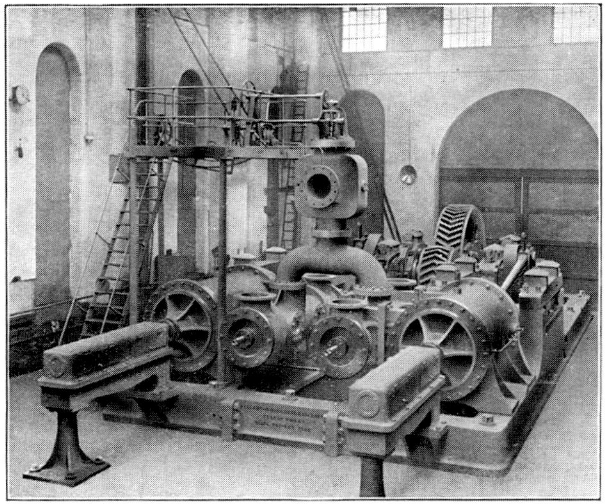

PAIR OF HORIZONTAL HIGH-PRESSURE ROLLING MILL ENGINES,
made for Messrs. A. & J. Stewart & Clydesdale, Ltd.

Cylinders, two at 45-in. diameter by 5-ft. 0-in. stroke.

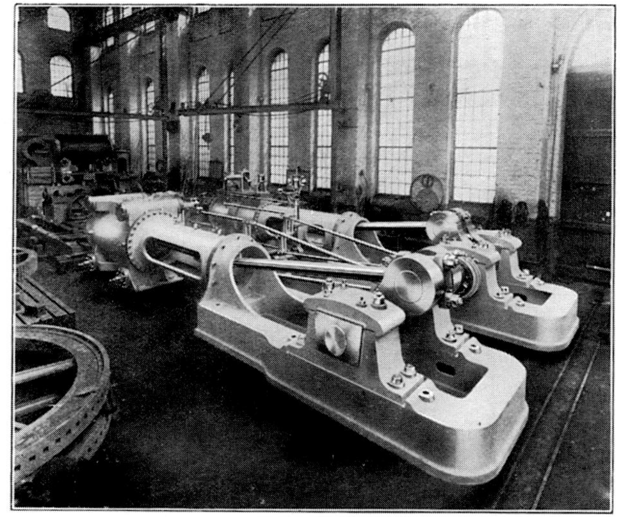

SIDE-BY-SIDE COMPOUND NON-CONDENSING ENGINES,
made for Messrs. Dick Kerr & Co., Ltd., Preston.

Cylinders, 23-in. and 40-in. diameter by 5-ft. 0-in. stroke.

FULLERTON, HODGART, & BARCLAY, LTD
PAISLEY

Fullerton

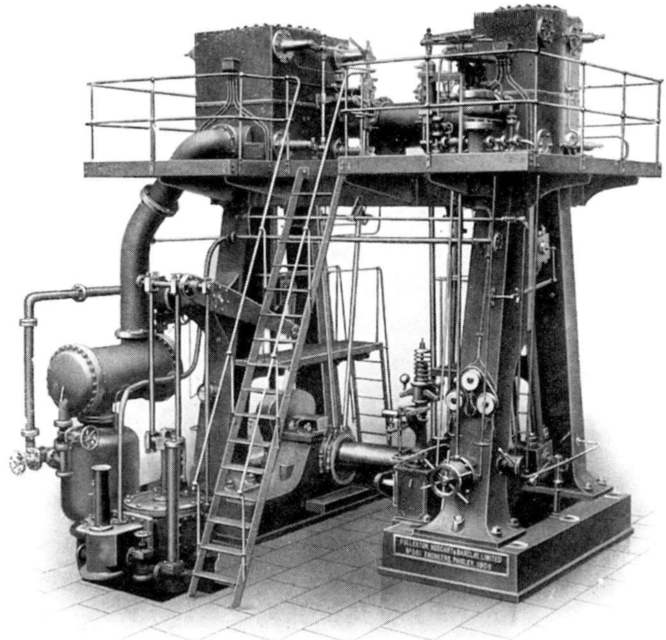

VERTICAL COMPOUND CONDENSING ENGINE.
Cylinders, 17-in. and 34-in. by 3-ft. 0-in. stroke.
160-lbs. per square in. 80 revs. per minute,
made for
The Craighall Milling Co. Ltd.

FULLERTON, HODGART, & BARCLAY, LTD
PAISLEY

MAK/132 - W. & J. Galloway, Knott Mill, Manchester.

1) Advertisements:
1867 (2), 1871, 1874, 1875, 1876, 1890, 1897 (see below), 1900 (see p.131), 1901, 1903, 1905, 1906, 1908, 1910 (2), 1921, plus two of no date.

Reverse: 1890 Bell's asbestos packing.

Reverse: 1897 Selig, Sonnenthal, 85 Queen Victoria Street, London - American machine tools; W. B. Brown Globe Works. Bankhall, Liverpool - wire ropes; J. & R. Houston, Cartsburn Foundry, Greenock - sugar machinery.

Reverse: 1900 W. H. Bailey & Co. Ltd, Albion Works, Salford, Manchester - steam and water valves.

Reverse: 1901 John Fowler & Co., (Leeds) - traction engines etc.

3) MS material:
Collection of MS notes by GW about the history and production of the firm. 12pp.

5) Printed material:
Photocopy of obituary of John Galloway (1804-1894) from *Engineering* 16 February 1894, with MS note by GW about output.

Collection of pages from *Proc. I.Mech.E*; Works visit to Knott Mill Works 19 May 1920; Obituary of John Galloway, 1894, with MS notes by GW.

Article from *Engineering* 17 August 1900, pp 203-210 description of the Galloway plant at the Paris Exhibition Electric Power Station.

Printed booklet: T. B. Morley, *The Extraction Steam Engine*, n.d. but c 1925.

Cutout folding advertisement for the Galloway patent oil valve gear, n.d. but c 1925.

6) Trade literature:
Booklet: *Uniflow steam engines*, 40 pp, n.d. but c 1922.

Booklet: *Rolling mill engines*, 34pp, n.d. but c 1920 (see p.130).

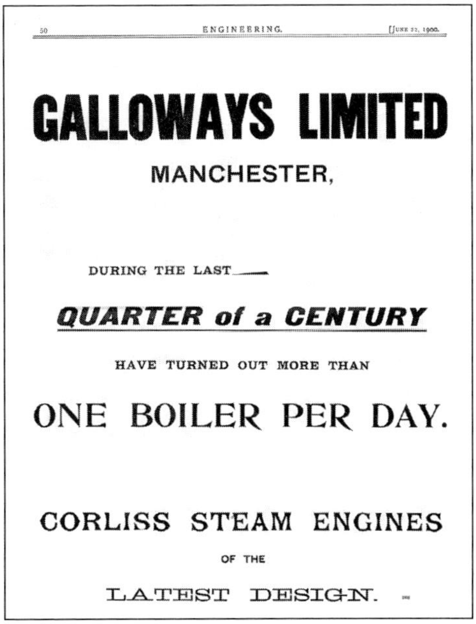

Piston Drop Valve Engines

Tandem Compound Type

The illustration (Fig. 21) shows a standard Galloway tandem compound piston drop valve engine, to work with superheated steam at 160 lbs. per square inch boiler pressure, with cylinders 17½in. and 33in. diameter, 3ft. 6in. stroke.

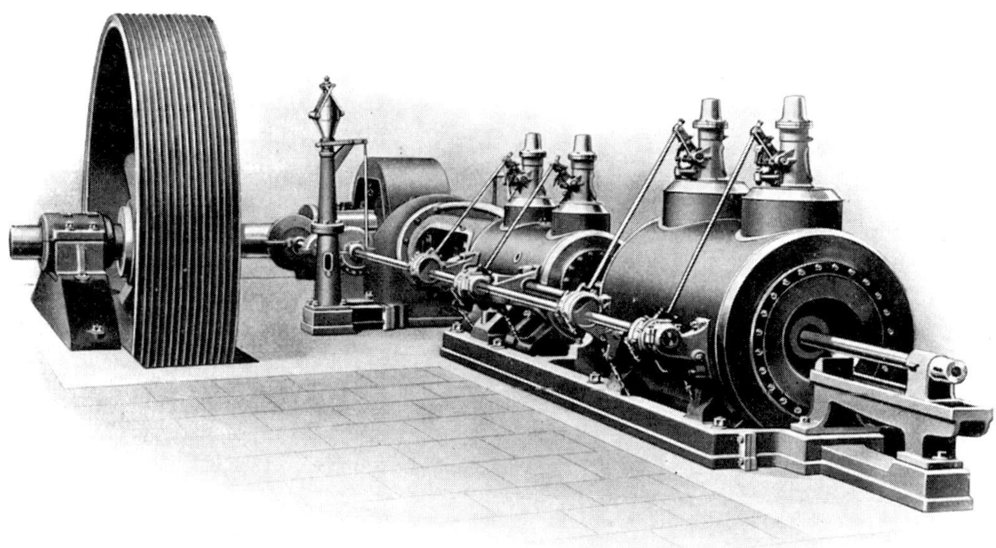

Fig. 21. Tandem Compound Piston Drop Valve Engine

This class of engine, which is a standardised production, will give the lowest possible steam consumption obtainable with a compound type of engine.

The flywheel is arranged with grooves for driving a continuous-running rolling-mill by means of ropes.

Several engines of this type have been supplied for driving tin-plate and bar mills, and they have given extremely economical working results.

MAK/133 - R. Gardner & Son, Novelty Works, Montreal, Canada.

2) Letters:
Letter to GW from Vapor Car Heating Co., 65 Dalhousie Street, Montreal, Canada, dated 22 September 1950 about the catalogue noted below.

6) Trade literature:
Illustrated catalogue of steam engines and baking machinery, 1876, 22pp [4 pages cut out].

MAK/134 - Richard Garrett, Leiston, Suffolk.

5) Printed material:
Article from *The Engineer* 20 March 1914 pp 309 - 312 describing a cement works at Claydon, Ipswich, with account of Garrett steam plant there. (see below).

6) Trade literature:
Catalogue of Locomotive Boilers, April 1925, 16 pp.

COMBINED SUPERHEATER ENGINES AND BOILERS

RICHARD GARRETT AND SONS, LIMITED, LEISTON, ENGINEERS

(*For description see page 319*)

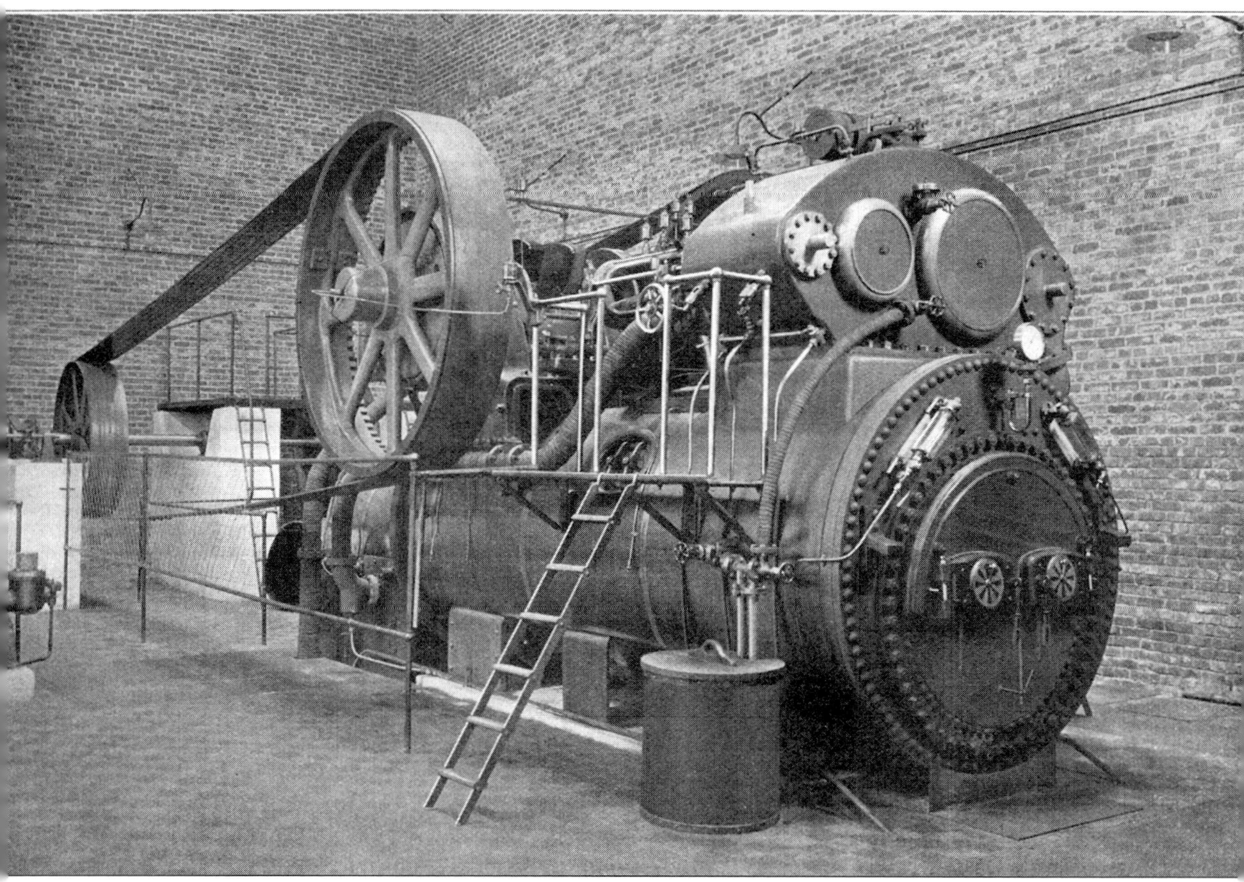

MAK/135 - John D. Gauldie, Broomhall Engine Works, West Scotland Street, Glasgow.

1) Advertisements:
1887, 1888.

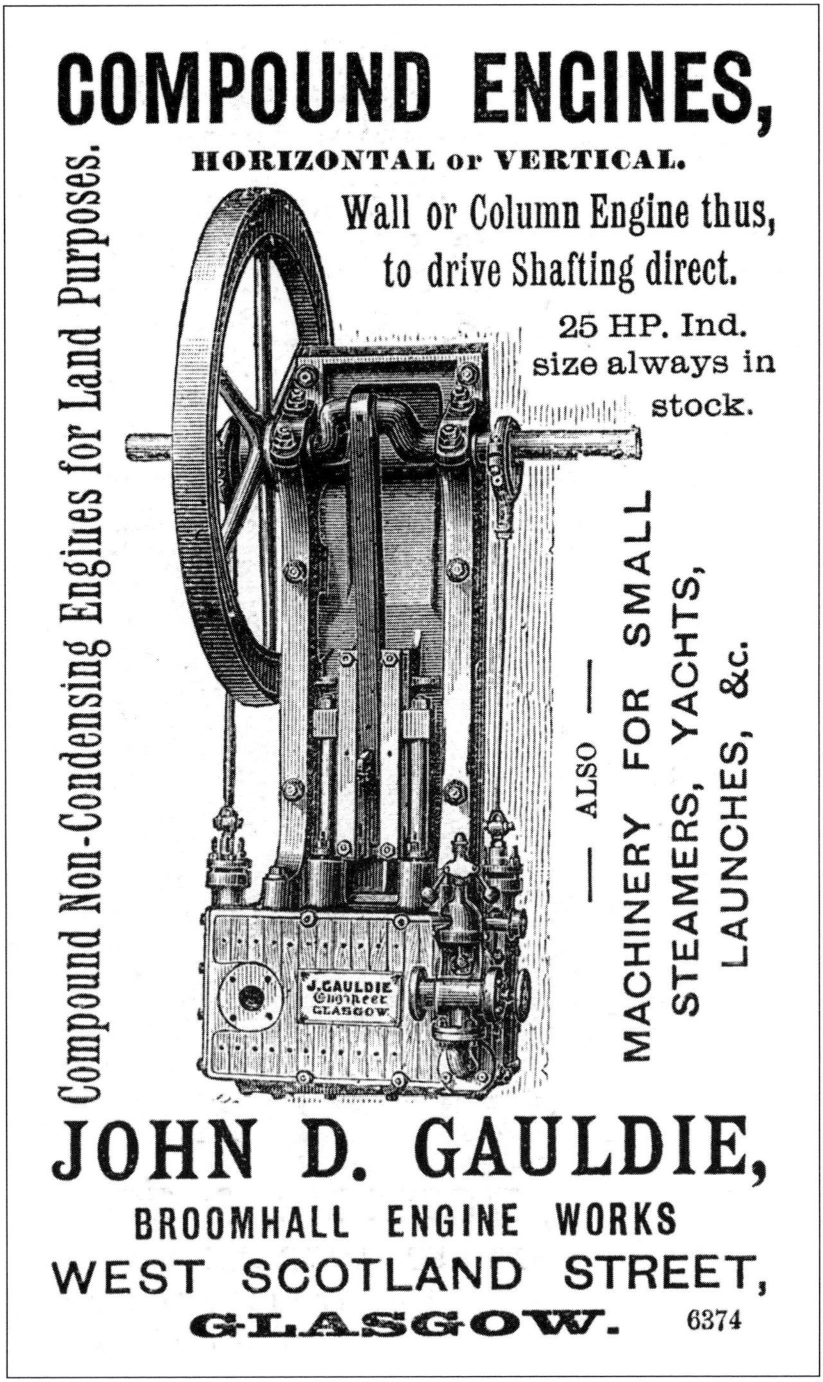

MAK/136 - Gimson & Co., Leicester.

2) Letters:
1) Letter to GW from Gimsons dated 18 November 1981 about GW visiting the works to examine the records.

3) MS material:
Collection of extracts from Gimson's order books listing boilers made 1895-1925.

MS notes by GW about Gimsons, 4pp. Notes there are over 160 list books of over 1000 pages each from 1885 - 1970; Highbridge, Weymouth, West Gloucestershire Water Works.

MAK/137 - Glenfield & Kennedy Ltd., Kilmarnock.

4) Photographs:
5 photographs of engines, mostly unidentified; two shown, below and opposite.

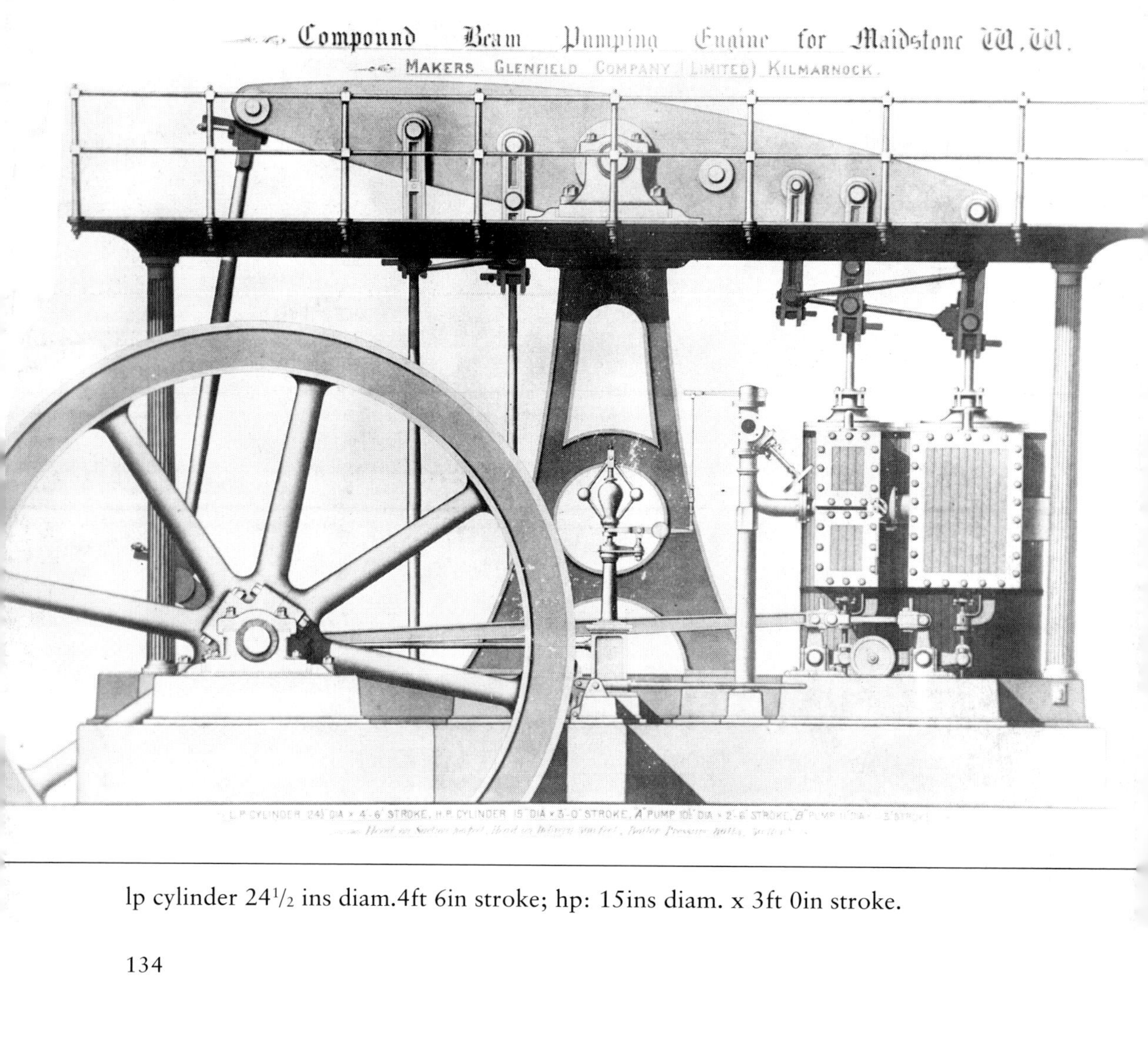

lp cylinder $24^1/_2$ ins diam.4ft 6in stroke; hp: 15ins diam. x 3ft 0in stroke.

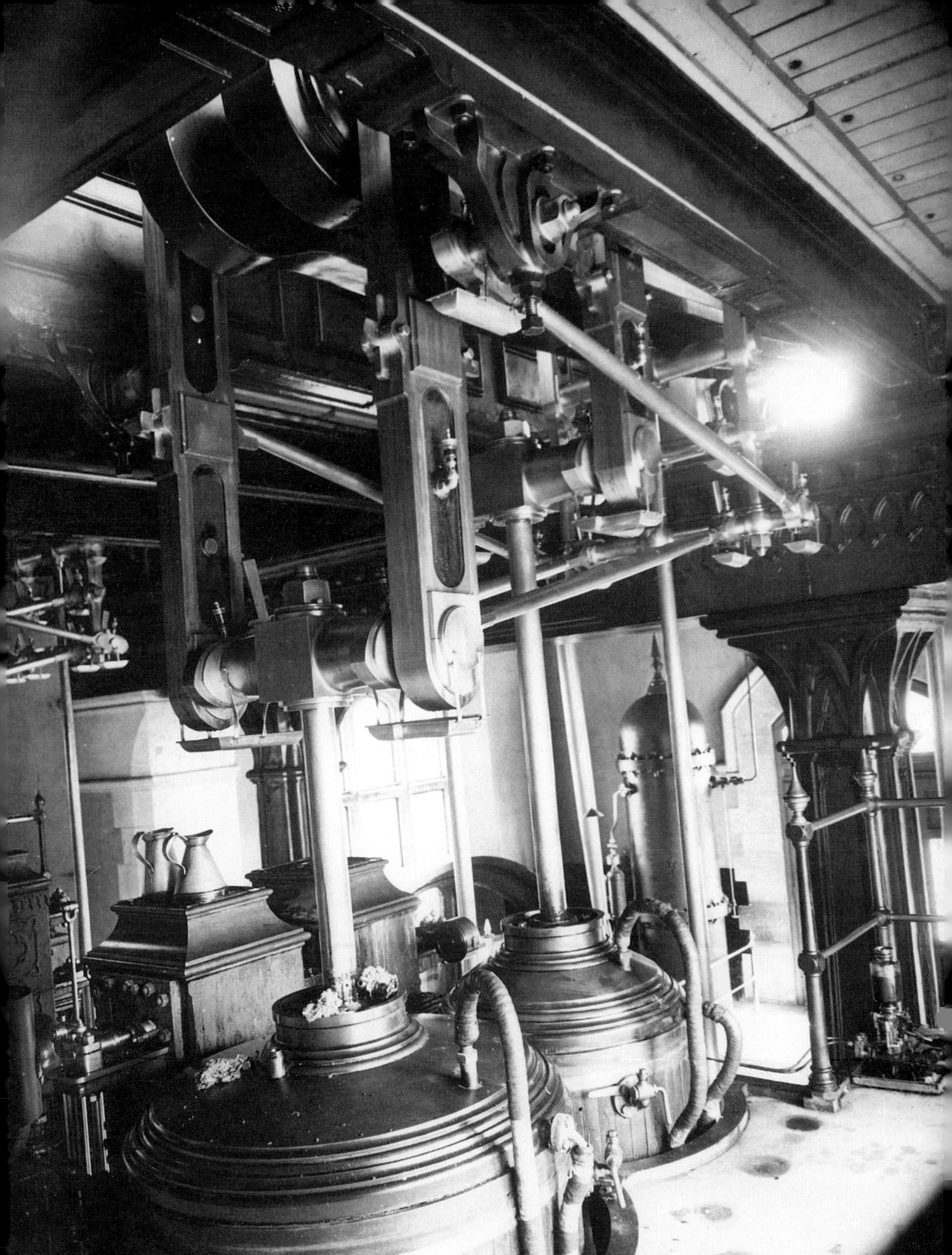

Blagdon Pumping Station, Bristol, 1902 (?).

MAK/138 - Globe Engineering Co., 38 Victoria Buildings, Manchester.

1) Advertisements:
1894.

[Oct. 26, 1894.

THE GLOBE ENGINEERING CO
LIMITED
38 VICTORIA BUILDINGS
MANCHESTER, ENGLAND
TELEGRAPHIC ADDESS "GLOBULUS MANCHESTER"

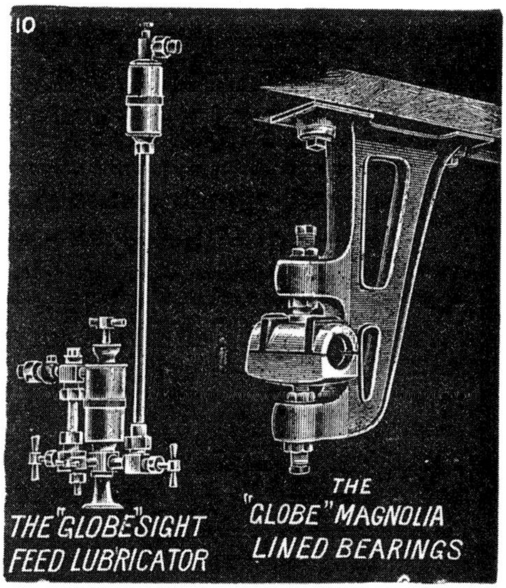

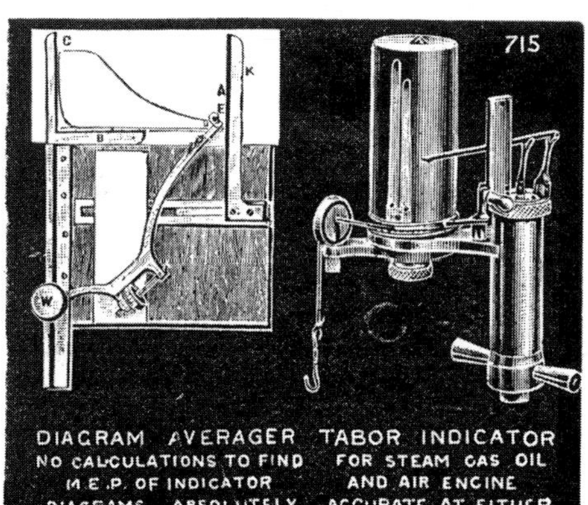

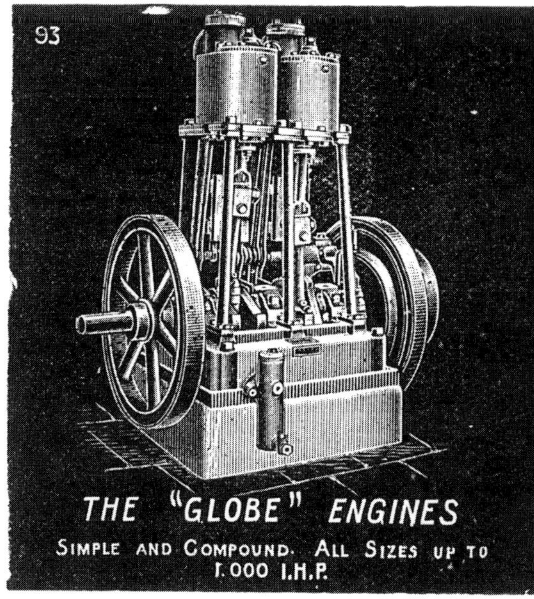

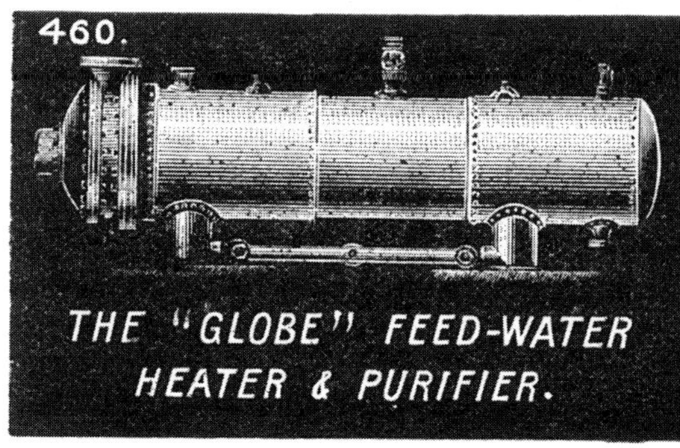

THE GLOBE ENGINEERING CO
LIMITED
38 VICTORIA BUILDINGS
MANCHESTER, ENGLAND
TELEGRAPHIC ADDESS "GLOBULUS MANCHESTER"

MAK/139 - Benjamin Goodfellow, Hyde, Cheshire.

1) Advertisements:
Page from unidentified and undated book with engraving of the works.

Ad from *Engineering* 23 October 1903 about auction of the works.

1890, 1893, 1894 (2) (see p. 139), 1895 (2), 1903.

Reverse: 1890 F. Reddaway, Pendleton, Manchester, - cotton belting; Peter Brotherhood, London; W. H. Allen, Lambeth.

Reverse: 1893, Alex Wilson, Vauxhall Ironworks, Wandsworth, London - duplex pumps.

Reverse: 1894 (both) Camerons, Manchester - plate bending rolls.

Reverse: 1895 (both) Thomas Piggot & Co., Atlas works, Birmingham - boilers & pipes.

Reverse: 1903 James Howden, Scotland Street, Glasgow - turbines and engines.

2) Letters:
Letter from J. Taylor to GW dated 12 April 1958.

Letter from Teddy Atherton to GW dated 14 April 1958.

Letter from Frank Wightman to GW, undated and incomplete - deals with dispersal of Goodfellow's drawings.

3) MS material:
MS notes by GW about Goodfellows.

4) Photographs:
Photographs of model horizontal Corliss engine made by Tony Walshaw.

5) Printed material:
Plate from *Mechanical World,* 24 February 1883 of pair of compound Corliss engines (2 copies).

Plate from *Mechanical World*, 24 March 1883 with more drawings of the compound Corliss engines. (see p. 138). For J. Mayall, Mossley.

Article from *Mechanical World,* 10 September 1887 describing horizontal Corliss compound engines for Astley Mill Co., Dukinfield.

Article from *Mechanical World*, 19 July 1890 describing vertical compound made for the Mannesman's tube works at Landore.

7) Other material:
Original Goodfellow drawings:

Dashpot and valve gear from order 417.

Manchester Corporation Electric Lighting Department. Rope wheel from order 2322.

Corliss engine for City of Leeds Tramways. Full size name plate for order 4954.

No. 2

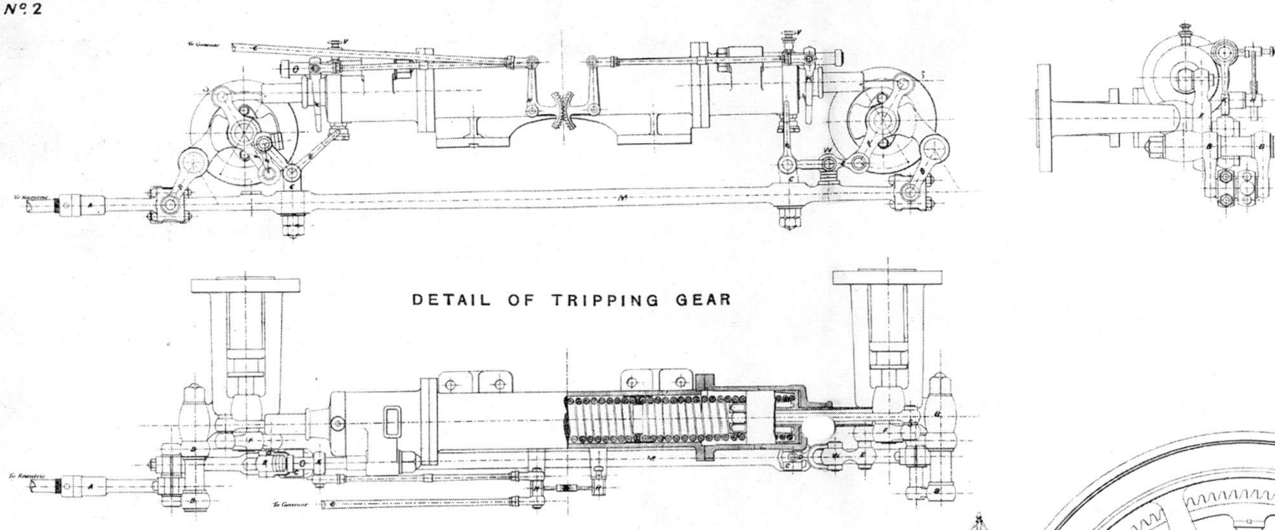

Pair of Compound Corliss Engines,

FITTED WITH RAMSBOTTOM'S PATENT TRIPPING GEAR.

Constructed to Indicate 1,100 H.P. when running at 36 revolutions per minute.

BY BENJAMIN GOODFELLOW, ENGINEER, HYDE,

FOR JOHN MAYALL, ESQ., MOSSLEY.

Cylinders, 26 inches and 52 inches diameter; 6 feet stroke.

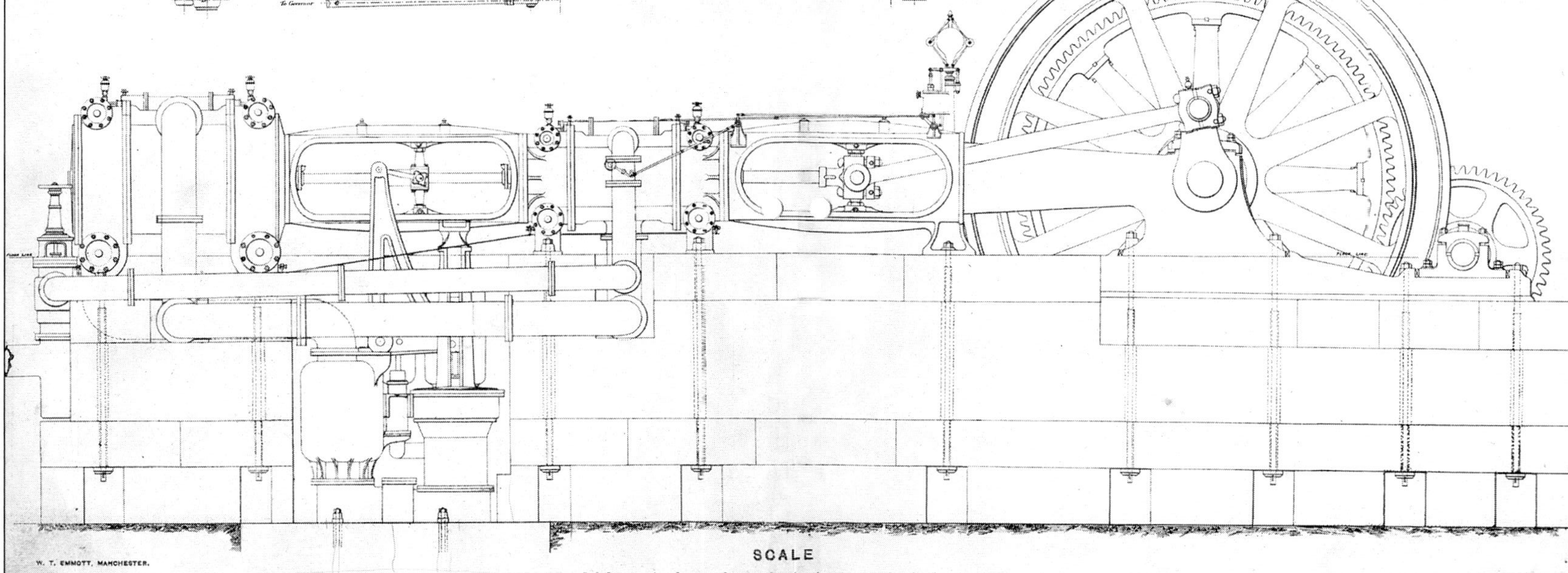

W. T. EMMOTT, MANCHESTER.

ESTABLISHED 1838. **BENJ[N.] GOODFELLOW,** ESTABLISHED 1838.

HYDE, NEAR MANCHESTER.

CONTRACTOR TO H.M. AND FOREIGN GOVERNMENTS.

MAKER OF **ROPE & WHEEL GEARING**

ALL DESCRIPTIONS OF MILLWRIGHT WORK.

HIGH-CLASS

STATIONARY PUMPING AND BLOWING ENGINES, with Corliss or other Valve Gears.

Improved METALLIC PISTONS & AIR PUMP BUCKETS.

Cylinders, &c., Bored, and Valve Facings Planed in their places.

Wheels, &c., Moulded by Patent Machinery.

9995

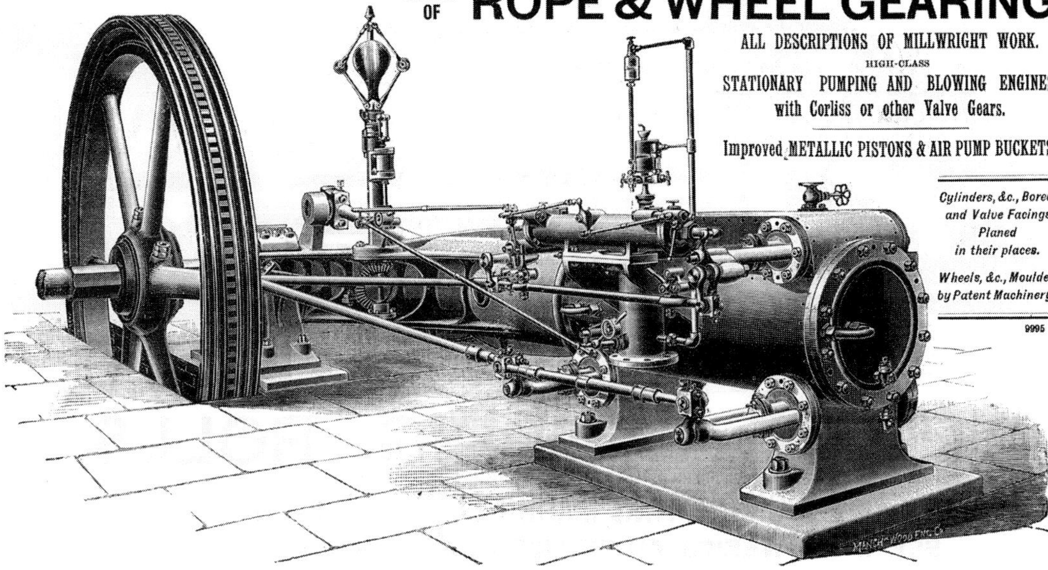

BENJ[N.] GOODFELLOW,

HYDE, NEAR MANCHESTER.

CONTRACTOR TO H.M. AND FOREIGN GOVERNMENTS.

Established 1838.

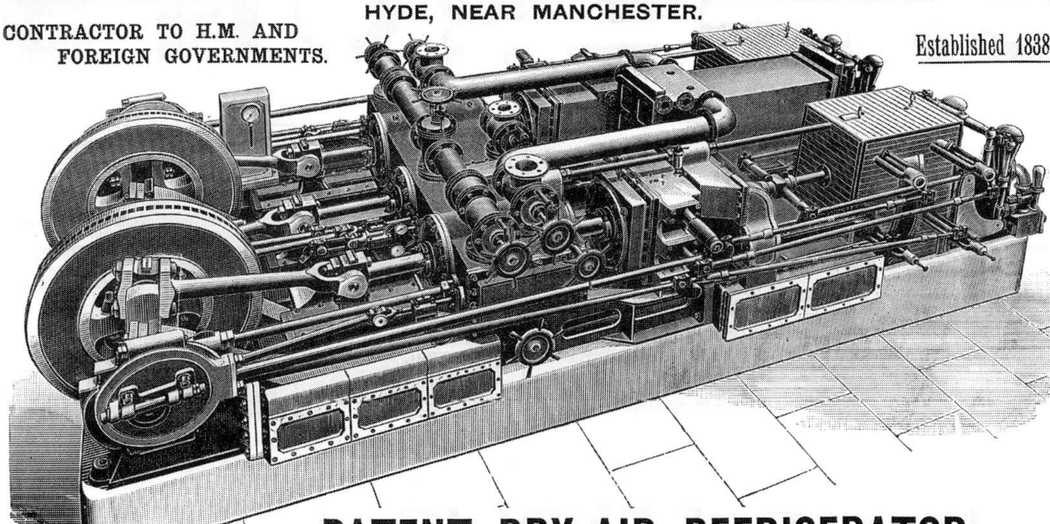

PATENT DRY-AIR REFRIGERATOR.

HIGH-CLASS CORLISS ENGINES,

For Pumping, Blowing, or any Stationary Purposes. Either Two or Multiple Cylinder Compound. 9995

Cylinders, &c., Bored, & Valve Facings Planed in their places. Wheels, &c., Moulded by Patent Machinery. Fly-wheels, Rope & Belt Pulleys, &c., turned to any size.

IMPROVED METALLIC PISTONS AND AIR PUMP BUCKETS.

MAK/140 - The Grantham Crank & Iron Co., Ltd, Grantham, Lincs.

1) Advertisements:
1888.

6) Trade literature:
Catalogue: No 26 of steam boilers, 58pp n.d. but c 1914.

MAK/141 - Green, Atkinson & Co., Leeds.

2) Letters:
Letter from Leeds Reference Library to GW dated 14 July 1978 saying they cannot trace anything about this firm.

MAK/142 - Greenwood & Batley, Albion Works, Leeds.

1) Advertisements:
1879, 1888, 1890 (with reduced copy), 1896, 1910, plus one with no date.

Reverse 1879 Edwin Lewis & Son, Wolverhampton - Iron boiler tubes.

Reverse 1888 Lancaster & Tonge, Pendleton - Piston rings.

Reverse 1896 Ropeways Syndicate, 150 Leadenhall Street, London - Aerial ropeways.

2) Letters:
Letter from Greenwood & Batley to GW dated 5 October 1951 with list of photographs 1885-1896 (see below).

Letter from Elton Cop Dyeing Co. to GW dated 31 March 1959 about steam turbines at the works.

Letter from Greenwood & Batley to GW dated 16 March 1959 with list of De Laval steam turbines with belt or rope drives for mill driving. Mentions sending drawings.

Letter from Greenwood & Batley to GW dated 25 March 1959 about sending more drawings and GW visiting the works.

Letter from Cressbrook Doubling Co. Ltd, nr Buxton to GW dated 1 April 1959 about steam turbines.

4) Photographs:
Collection of photographs 1885-1896 referred to in letter of 5 October 1951.

[Note: these are not complete since several listed in the letter are missing and there are two not listed]. (see pp. 141-142).

5) Printed material:
Article from *Mechanical World* 25 January 1895, pp 36, 37 describing 400 hp compound mill engine, with photocopy of p 37.

7) Other material:
Dyeline print of 329 hp De Laval steam turbine at Elton Cop Dyeing Co. Ltd.
Dyeline print of 450 hp turbine and condensing plant at J. Bright Bros, Rochdale.

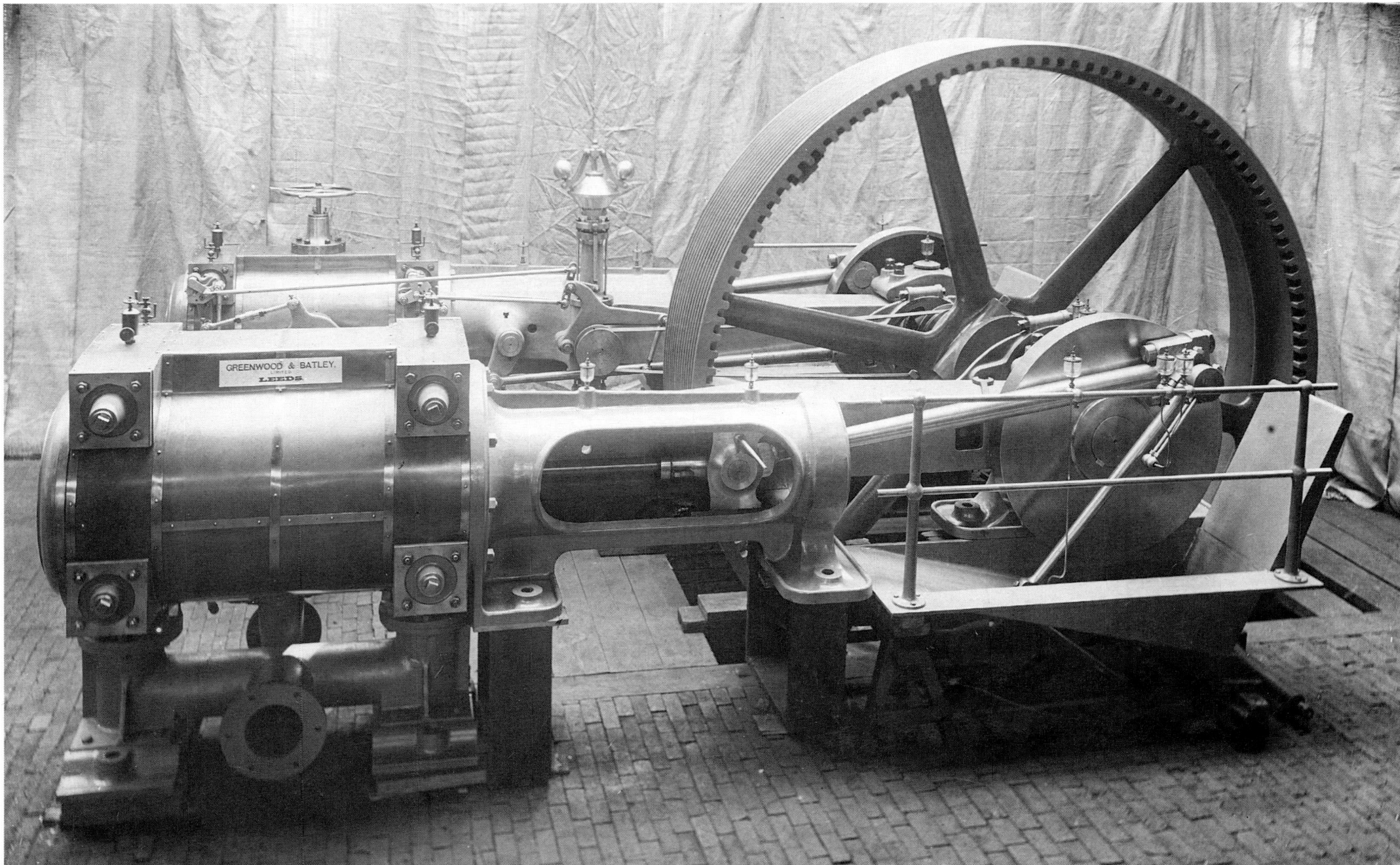
GREENWOOD & BATLEY,
LEEDS.

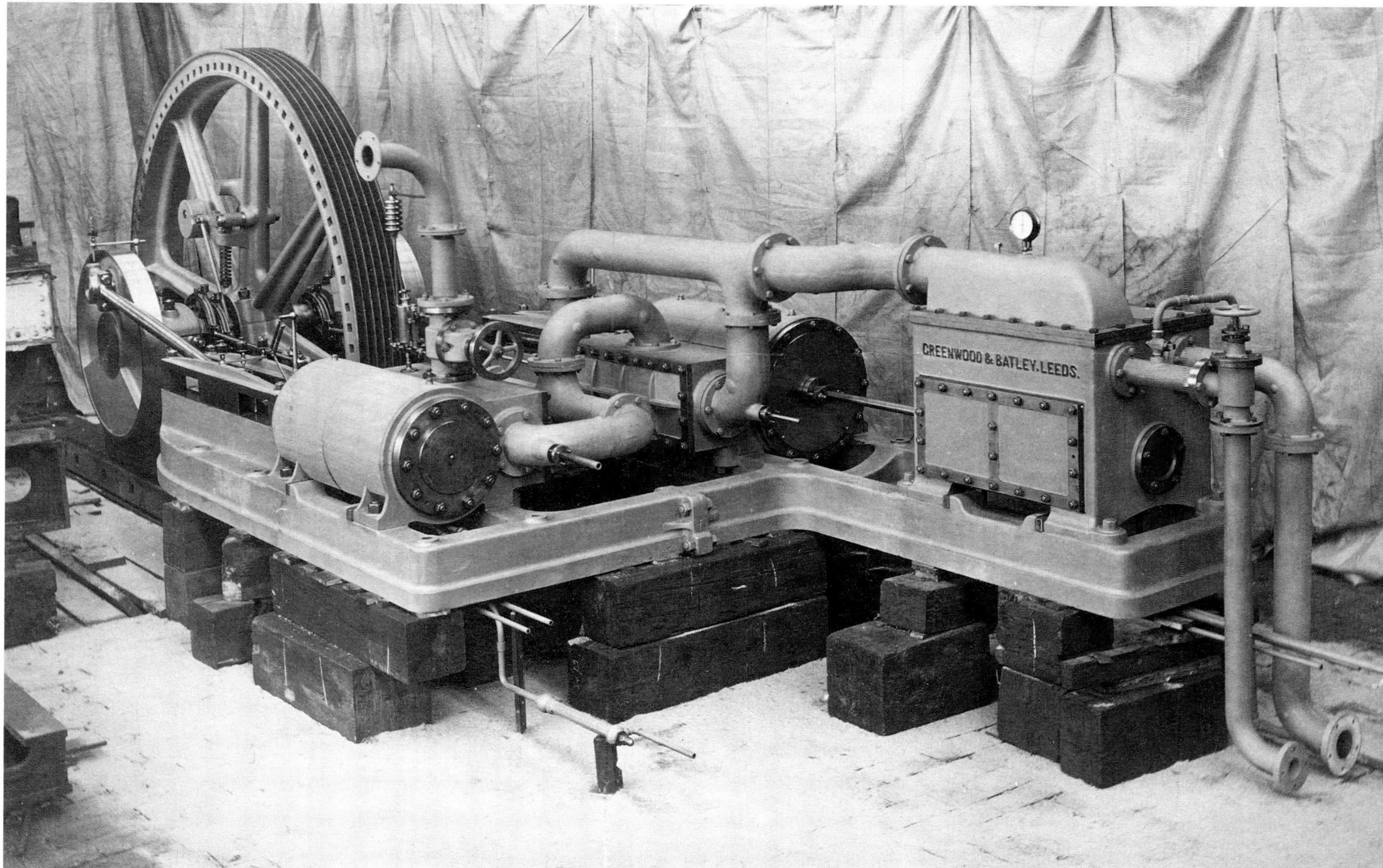
GREENWOOD & BATLEY. LEEDS.

MAK/143 - Gutehoffnüngshutte, Sterkrade, Germany.

5) Printed material:
Article from *Engineering* 8 August 1902 about 3000 hp triple expansion engine.

MAK/144 - Guy, Swift & Co., Engineers, London & Thrapston.

1) Advertisements:
1875.

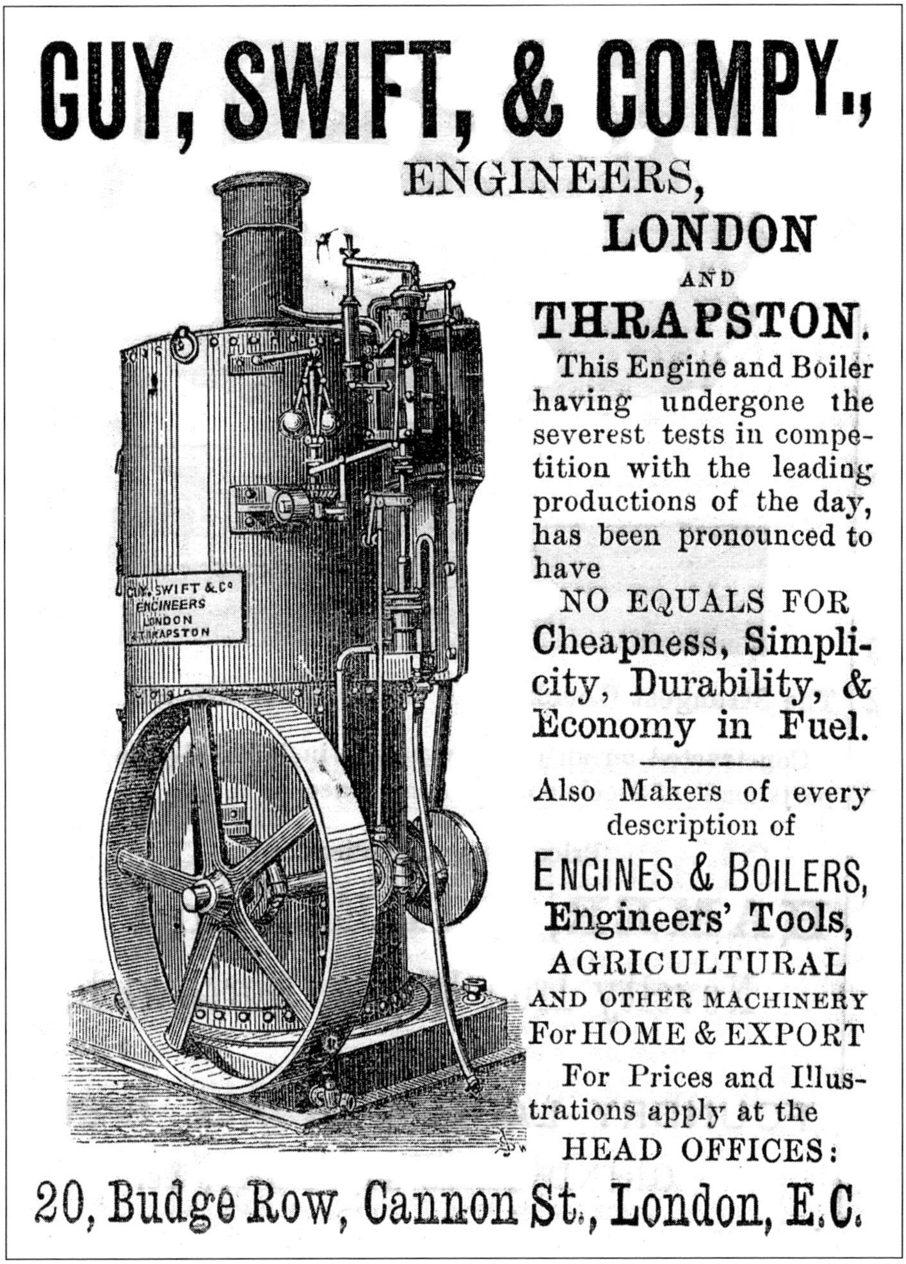

MAK/145 - J. & M. Gwynne, Hammersmith Ironworks, London.

1) Advertisements:
Ads for 1874, 1879, 1901(2).

> Reverse 1874 Girdwood, East India Dock Road, Poplar, London - patent metal packing; A. Verey, Buckland Engine Works, Dover - launch and yacht engines; Richard C. Gibbins & Co., Berkeley Street, Birmingham - hand and machine cranes; Davey Paxman, Colchester - vertical engine & boiler.

2) Letters:
Letter from Ailen Gwynnes Pumps to GW dated 29 June 1977 saying certain records had been destroyed.

3) MS material:
MS notes by GW on the history of the firm, 2pp.

MAK/146 - J. W. Hackworth, Darlington Engine Works, Darlington.

1) Advertisements:
1876.

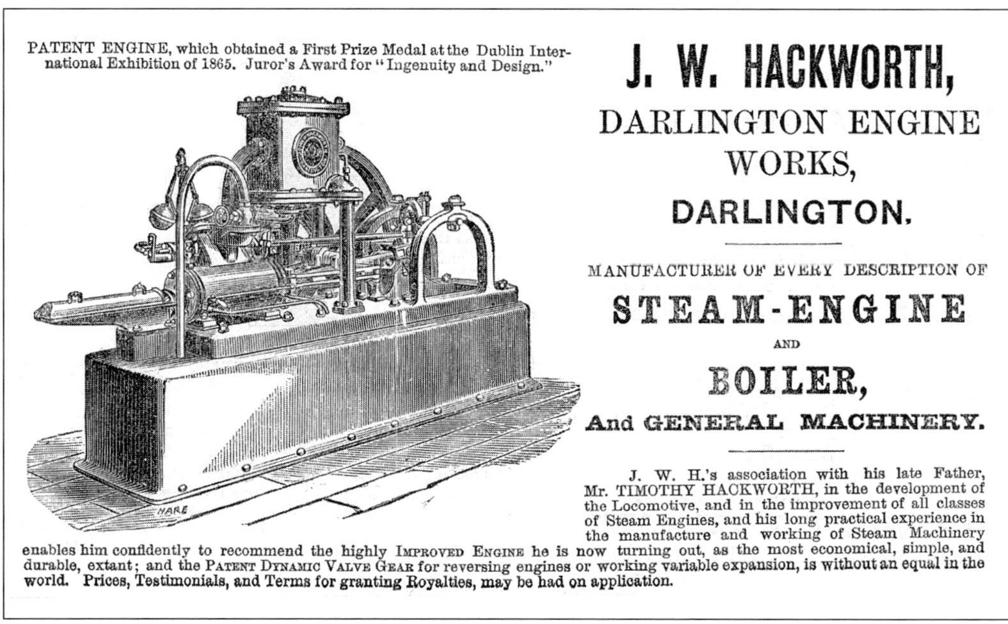

MAK/147 - Haigh Ironworks, Liverpool.

2) Letters:
Letter from Liverpool Record Office to GW dated 11 November 1977 listing directory references.

MAK/148 - J. & E. Hall, Dartford, Kent.

1) Advertisements:
1893.

2) Letters:
Letter from J. & E. Hall to GW 11 May 1948 about engines at Nicholson's gin distillery, Bromley-by-Bow.

4) Photographs:
6 photographs - 3 of old beam engines and 3 steam driven CO_2 compressors.
Below: Tonge Mill, Sittingbourne (Wicks Brothers). P. 146: Unknown location.

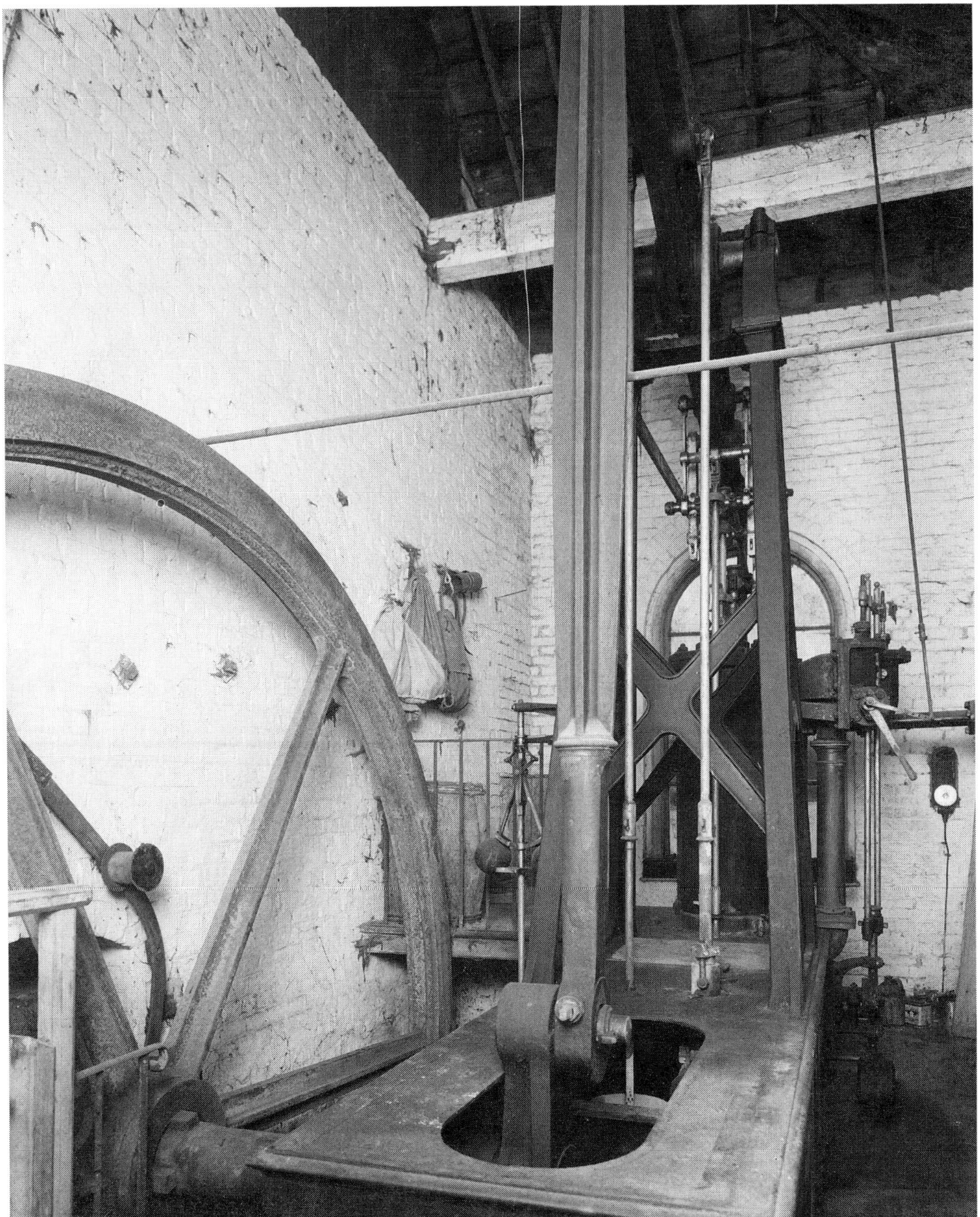

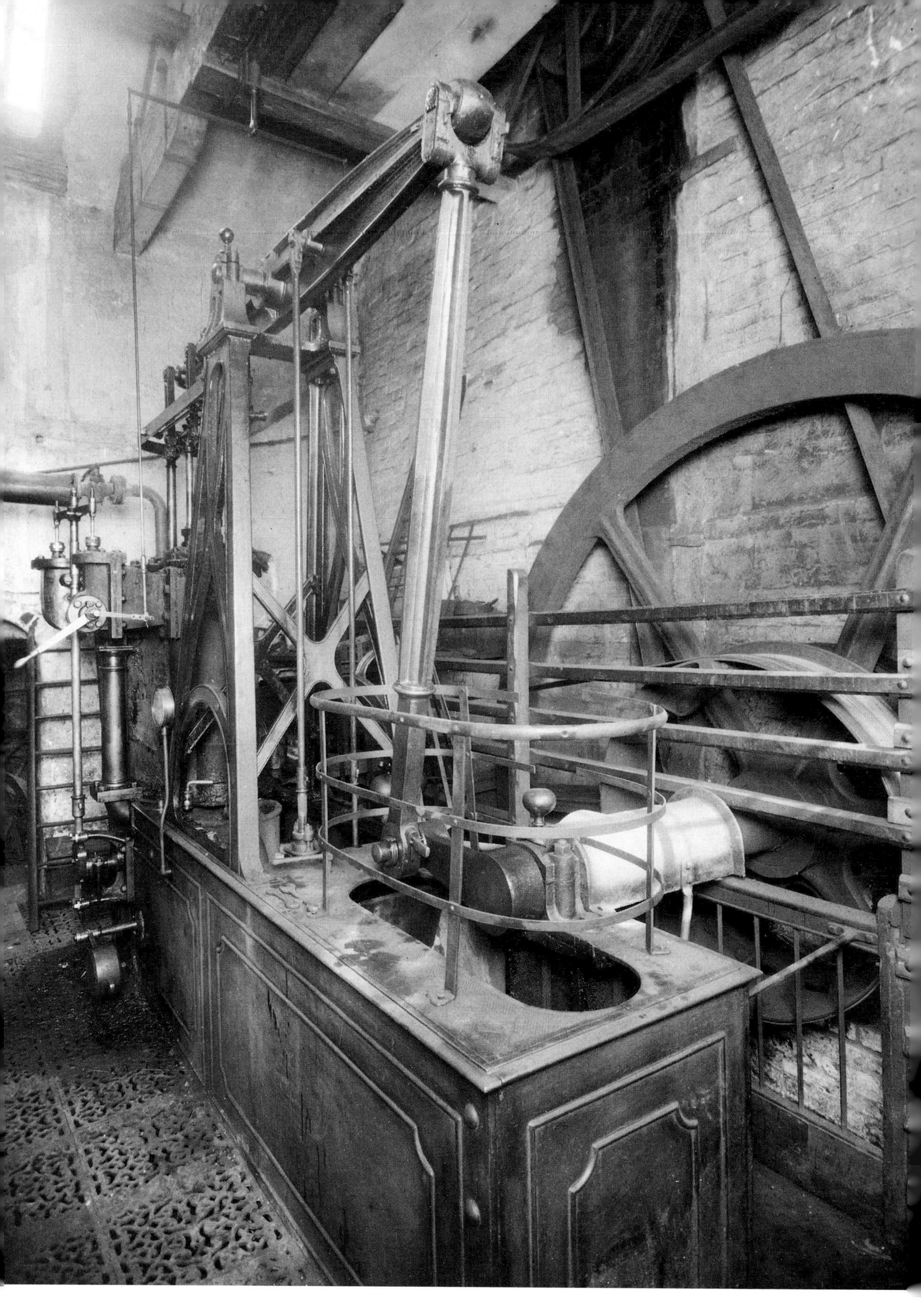

MAK/149 - J. P. Hall, Oldham.

1) Advertisements:
1903 (*Engineering*; see below).
Reverse: 1903 British Thomson-Houston - electric motors.

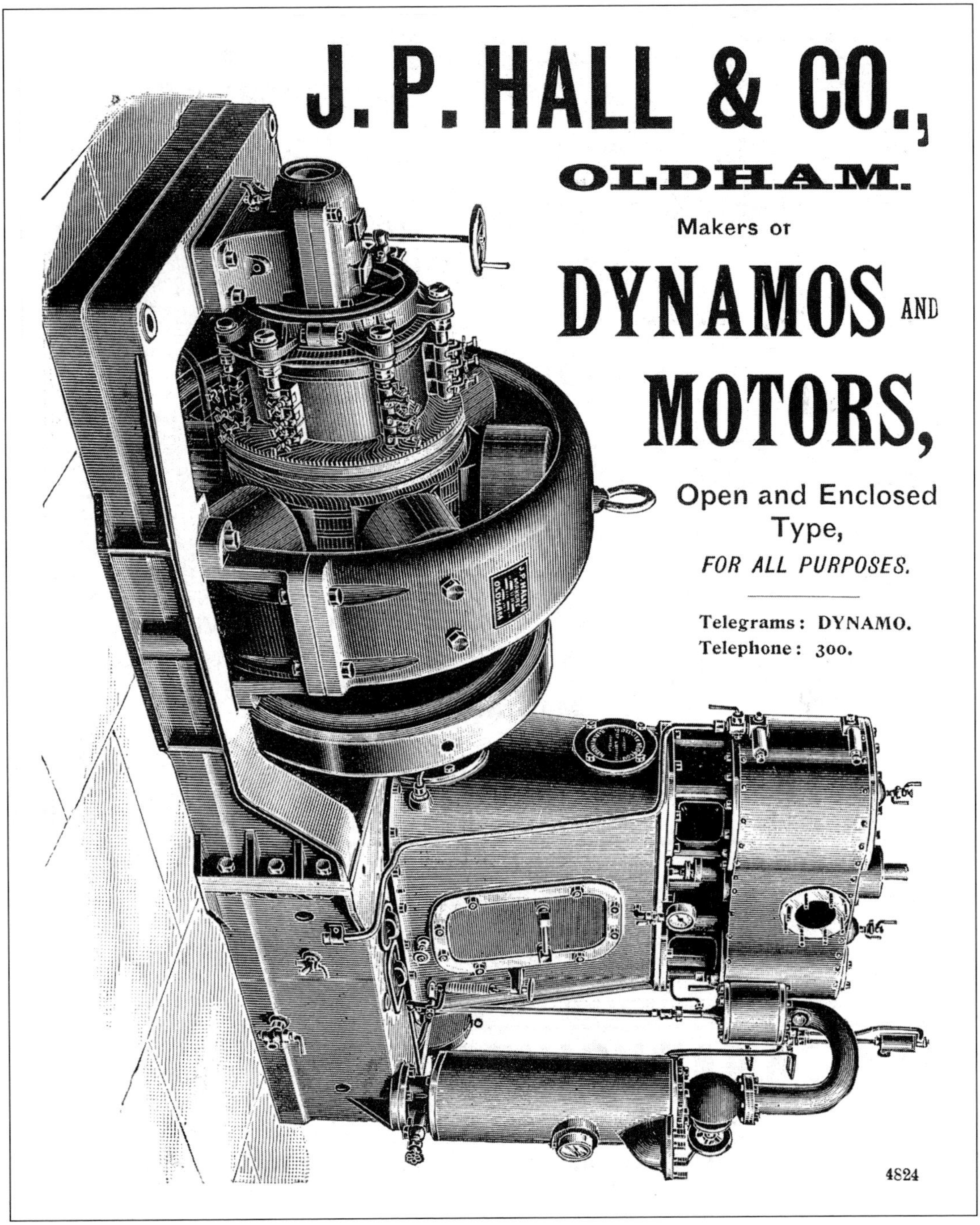

MAK/150 - Haniel & Lueg, Düsseldorf - Grafenberg.

5) Printed material:
Article from *Engineering*, 11 July 1902, p 57 about electric mine-draining plant at the Düsseldorf Exhibition.

MAK/151 - J. Hargreaves & Co., Leebrook Foundry, Rawtenstall.

5) Printed Material:
Article from *Mechanical World* 26 July 1895 about compound mill engine for a felt carpet works.

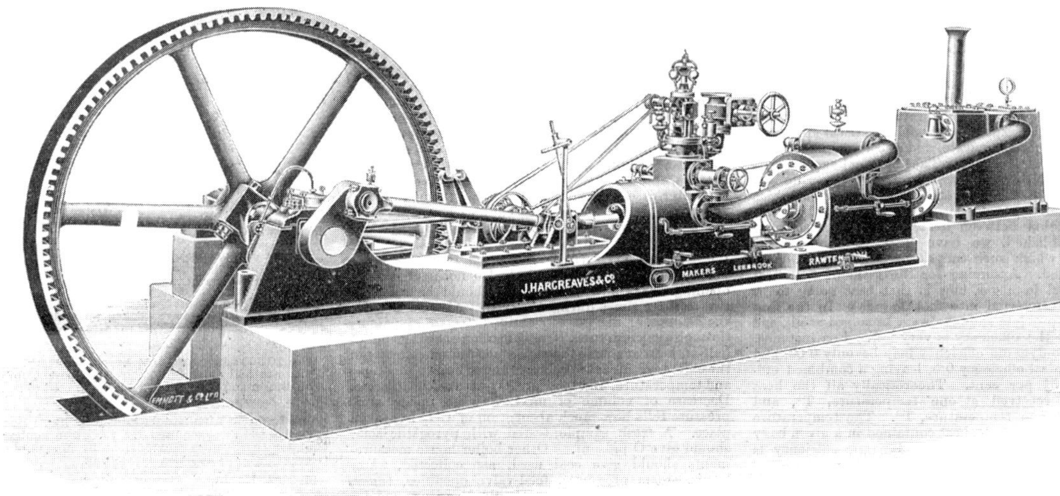

Compound Mill Engine.

MAK/152 - Harland & Wolff Ltd., Belfast.

2) Letters:
Letter from Harland & Wolff to GW dated 24 February 1954 with details of specific engines, also that their records were destroyed by bombs.

MAK/153 - Harris, Providence, R. I., USA.

15) Printed material:
Photocopy from *Power* June 1902, p5 illustrating a 10 hp cross-compound Harris - Corliss engine.

MAK/154 - Harrisburg Foundry & Machine Works, Harrisburg, USA.

15) Printed material:
Article from *The Engineer,* 11 November 1894, describing 600 hp tandem compound horizontal engine (not complete).

MAK/155 - Hartley & Sugden, Halifax.

6) Trade literature:
Catalogue of central heating boilers, 1951, 166 pp.

MAK/156 - Harvey & Co., Ltd, Hayle, Cornwall.

1) Advertisements:
1871, 1874, 1890, 1893 (see below), 1894, ads for sale of works 1894, 1904.

Same side: 1871 Richard Moreland & Son - steam roller

Reverse 1871 Bryan Donkin, Bermondsey - Gas Valves.

Reverse 1890 David Colville & Sons - steel makers; Weardale Iron and Coal Co., - boiler plates.

Reverse 1893 C. A. Parsons & Co., Heaton Works, Newcastle - turbines.

Reverse 1894 Burnham Brick & Lime Co. - Portland Cement; Scriven & Co., Leeds - machine tools.

5) Printed material:
Photocopy from *The Engineer* 1890, p 310 - 2 copies.

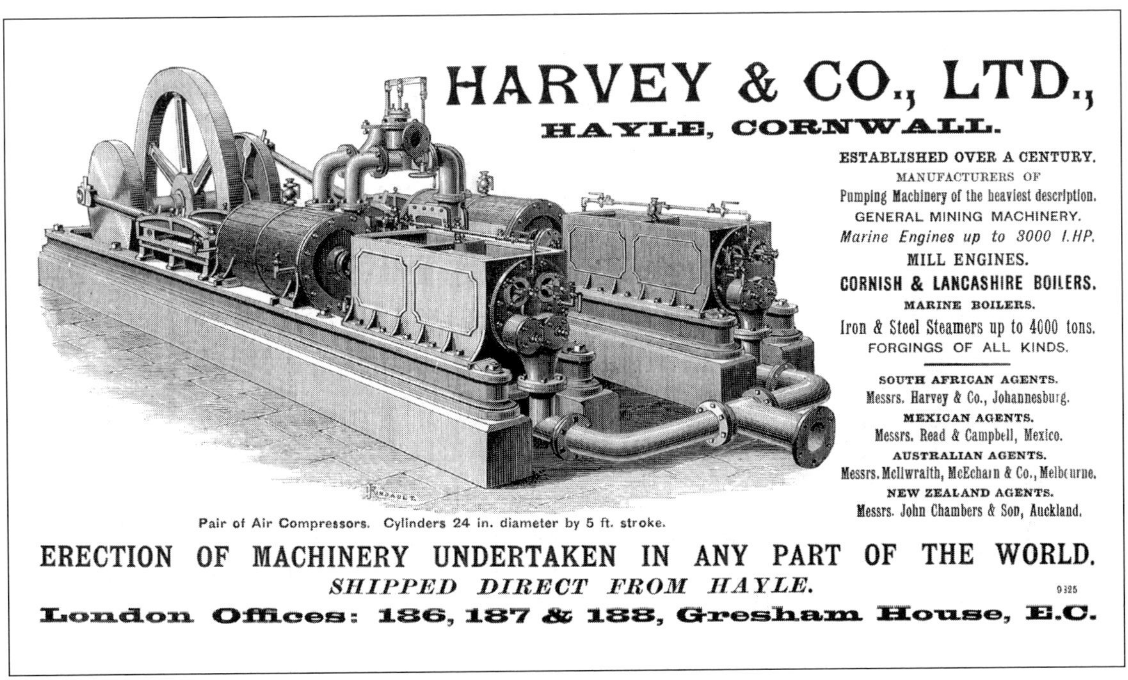

MAK/157 - Haslam Foundry & Engineering Co., Union Foundry, Derby.

1) Advertisements:

1879.

Reverse: 1879 Thwaites Brothers, Vulcan Ironworks, Bradford - a steam hammer; Slack's Emery Wheel and Machine Co. Ltd, Lime Bank Works, Ashton Old Road, Manchester; Savile Street Foundry & Engineering Co. Ltd, Sheffield. - stone-crushers and blowers.

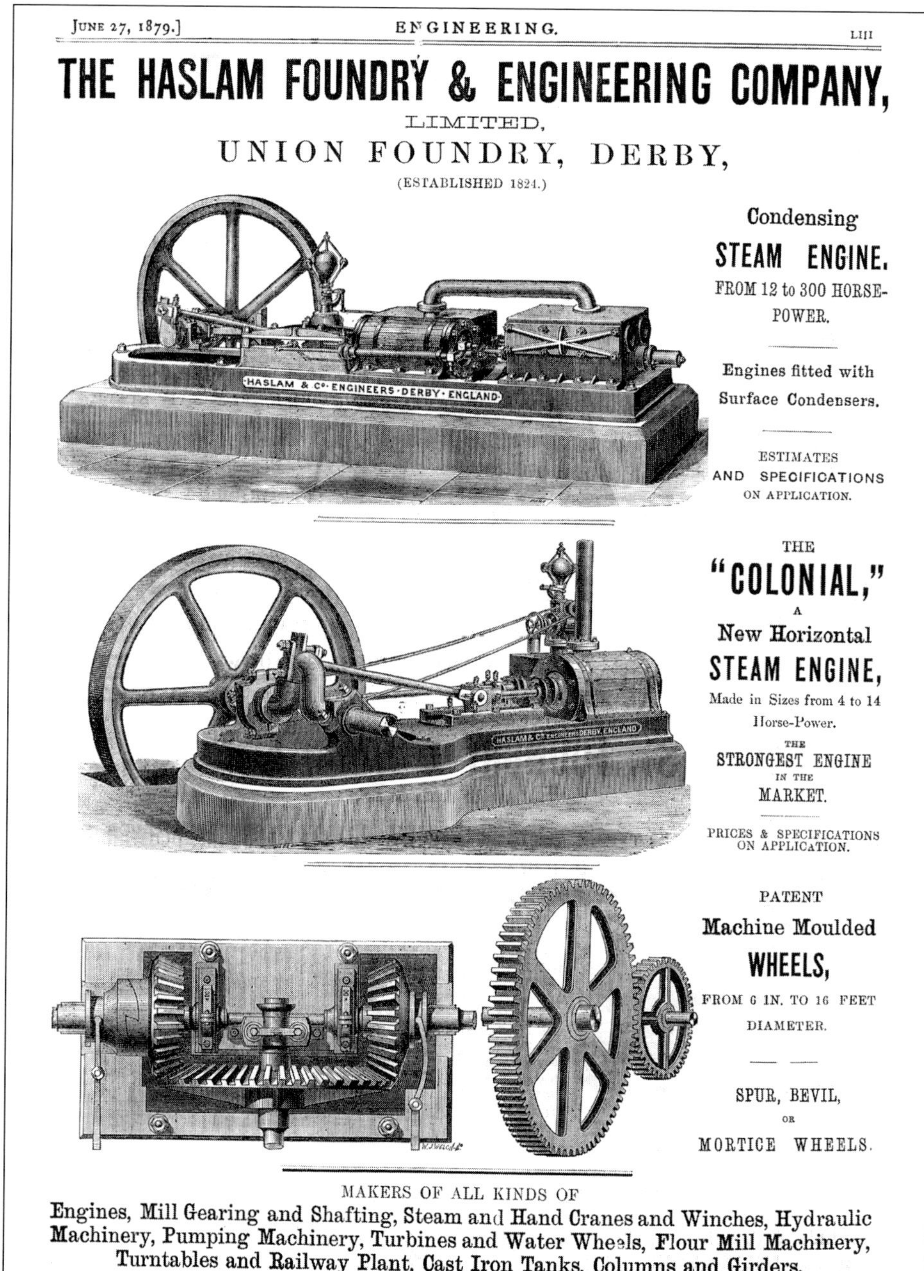

JUNE 27, 1879.] ENGINEERING. LIII

THE HASLAM FOUNDRY & ENGINEERING COMPANY,

LIMITED,

UNION FOUNDRY, DERBY,

(ESTABLISHED 1824.)

Condensing STEAM ENGINE. FROM 12 to 300 HORSE-POWER.

Engines fitted with Surface Condensers.

ESTIMATES AND SPECIFICATIONS ON APPLICATION.

THE "COLONIAL," A New Horizontal STEAM ENGINE, Made in Sizes from 4 to 14 Horse-Power. THE STRONGEST ENGINE IN THE MARKET.

PRICES & SPECIFICATIONS ON APPLICATION.

PATENT Machine Moulded WHEELS, FROM 6 IN. TO 16 FEET DIAMETER.

SPUR, BEVIL, OR MORTICE WHEELS.

MAKERS OF ALL KINDS OF

Engines, Mill Gearing and Shafting, Steam and Hand Cranes and Winches, Hydraulic Machinery, Pumping Machinery, Turbines and Water Wheels, Flour Mill Machinery, Turntables and Railway Plant, Cast Iron Tanks, Columns and Girders.

MAK/158 - Hathorn Davey & Co., Sun Foundry, Leeds.

1) Advertisements:
Ads 1894, 1901, 1903, 1905.

> Reverse 1894 William Johnson, Castleton Foundry, Armley, Leeds; Peter Brotherhood; W. H. Allen, Queen's Works, Bedford.

2) Letters:
Letter to R. A. Buchanan from Hathorn Davey, dated 15 May 1974 with detailed summary of the company records.

Carbon copy of letter to Hathorn Davey from GW in reply to the above, dated 18 June 1974.

3) MS material:
3 large envelopes of MS notes by GW about Hathorn Davey, with photocopies of pages from catalogues, stock books etc.

4) Photographs:
Large print of engine at Dancers End. (see p.152).

7) Other material:
Booklet: *Description of the Differential Expansive Pumping Engine* by Henry Davey, 1880, 36pp plus plates - inset leaflet about Davey's differential valve gear.
Typed history of the firm by Hugh Lupton, November 1940, 14pp.

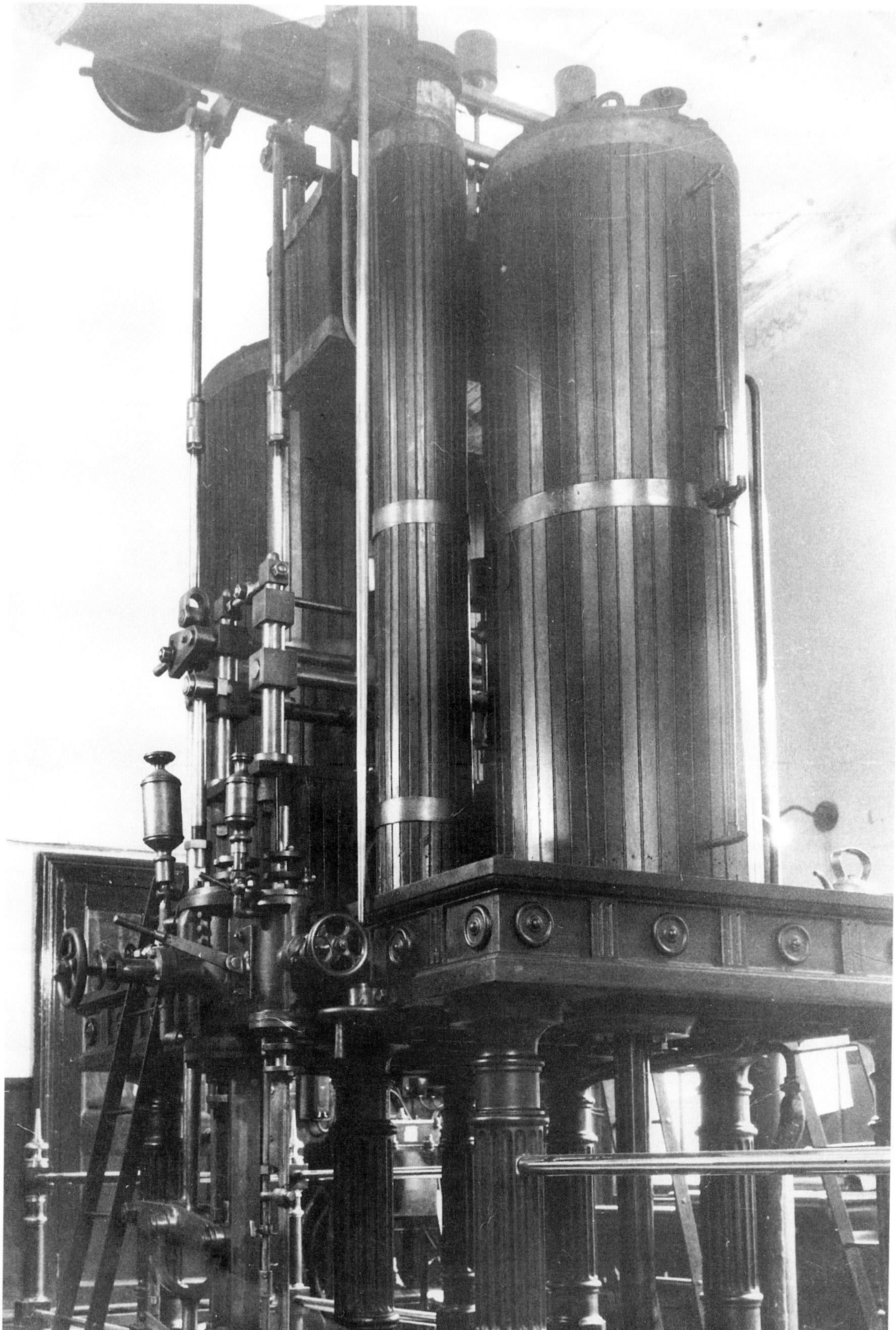

MAK/159 - Hatley Engine Co., Fairweather Green, Bradford.

6) Trade literature:
Photocopy of catalogue: nd but c 1914, 29pp.

MAK/160 - Hayward, Tyler, Luton, Beds, formerly of Whitecross Street, London.

1) Advertisements:
1867, 1879, 1892, 1894, 1896, 1901.

Reverse: 1867 Alexander Chaplin & Co., Cranston Hill Engine Works, Glasgow; Henry Kinsey, Robin Hood Works, Nottingham.

Reverse: 1879 Edwin Lewis & Sons, Wolverhampton – boiler tubes.

Reverse: 1894 Doulton & Co., Lambeth – plumbago crucibles.

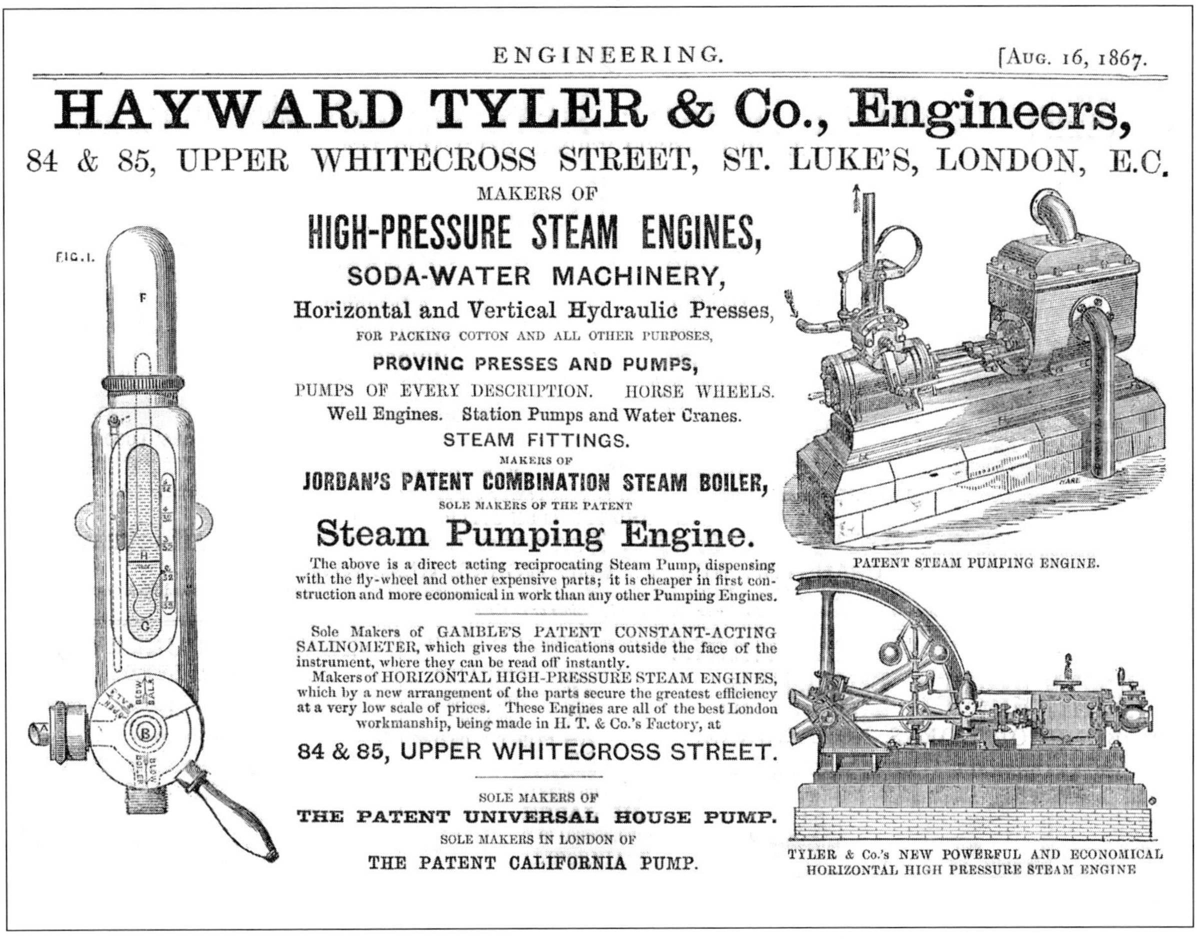

Opposite page: Hathorn Davy engine at Dancers End.

MAK/161 - K. W. Hedges, 36 King William Street, London Bridge.

1) Advertisements:
1875.

MAK/162 - Heenan & Froude, Manchester.

1) Advertisements:
1887.

MAK/163 - Helios Company, Cologne.

5) Printed material:
Article from *Engineering* 6 July 1900, p 11, describing 2000-volt generator at Paris Exhibition.

MAK/164 - Hick Hargreaves, Soho Ironworks, Bolton.

1) Advertisements:
1921.

2) Letters:
Letters from Hick, Hargreaves to GW dated 4 October 1957, 6 May 1959, with engine data.

Letter from Hick, Hargeaves, 13 May 1963 noting what the firm had done with the records of defunct firms they had taken over.

Letter from Ulster Museum to GW dated 4 May 1970 with photograph of unidentified beam engine.

Letter from GW to Ulster Museum dated 6 May 1970 confirming it was a Hick engine made 1880-84.

Letter with photograph from Harold Illingworth to GW dated 27 October 1972 - concerns 10 ft Corliss engine at Daniel Illingworth's Whitley Mill, Bradford. NB note on back of print about history.

Letter from Dave Collier to GW dated 20 May 1980 relating to a collection of copies of Hick photographs in the envelope, with reply from GW dated 24 May on reverse.

3) Ms material:
Collection of MS notes by GW about the firm - 8pp.

4) Photographs:
Collection of 15 works photographs in envelope marked 'Uniflow and drop valve' - no details of which engines or dates (one engine no. 919).

Collection of 23 works photographs in envelope labelled 'Corliss' - no details of engines or dates.

Collection of 31 works photographs in envelope labelled Diesel engines, shops & engine details.

Collection of 5 photographs in envelope labelled 'Hick Hargreaves, J. & E. Wood', courtesy of Dave Collier; Not dated or identified strip of 3 - 2 $^{1}/_{4}$ in sq negatives of catalogue - illustration of large cross-compound.

Print of beam engine at Lowes Mill, New Mills, Stockport [Egyptian design, like Marshall's engine at Leeds] (see p. 156), with loose sheet of dimension notes by GW. Science Museum print of works photograph of 10,000 HP verticals for Deptford Power Station.

5) Printed material:
Patent specifications No 2933 of 1881 (Valve gear) and 1741 of 1883 (Flywheels and pulleys).

Photocopy of plate from *The Engineer* 28 May 1869 of vertical Corliss engine at Saltaire.

Article from *The Builder* 3 October 1874 describing new cotton mill of Tootal, Broadhurst & Lee at Bolton - with Hick engines.

Article from *Mechanical World* 20 August 1895 describing engine at Eagley Mills, Bolton.

Article from *Mechanical World*, 27 July 1889 describing a vertical condensing Corliss engine.

Article from *Mechanical World*, 1 March 1890 describing a 4000 hp compound mill engine, with photocopy.

Article from *The Engineer* 11 January 1895 describing compound mill engine at Saltaire.

Article from *Mechanical World*, n.d., describing triple expansion mill engine, for Nippon Mills, Japan.

Article from *The Engineer* 19 July 1895 describing Leicester Corporation electricity works.

Article from *Engineering* 1 January 1897 describing combined engine and alternator at Bolton.

Article describing a portable non-condensing engine, from *Imperial Journal of Science*, c 1850.

Article from *Model Engineer* 31 March 1955 by W. Watson on Hick engines at unidentified Lancashire spinning mill.

6) Trade literature:
Photocopy of pages from unidentified catalogue:, c 1925, 19pp, lacks front page.

Photocopy of catalogue: *Steam engines* c 1927, 28pp.

7) Other material:
Booklet: *100 years of engineering progress*, 1933, 20 pp.

Tracing by ?Frank Wightman of barring engine, no date or details.

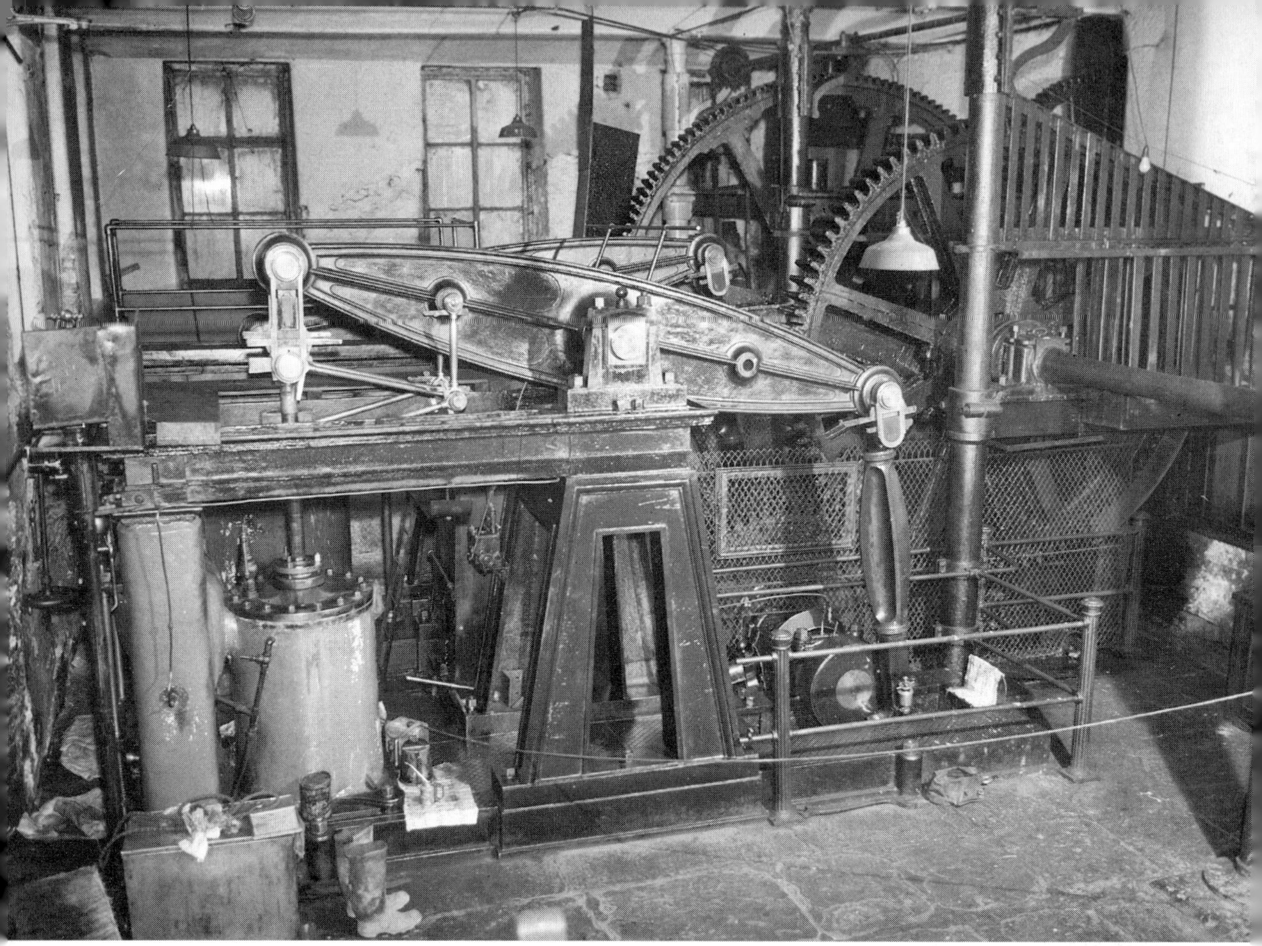

Above: Beam engine, Lowes Mill, New Mills near Stockport. Worked with a waterwheel.

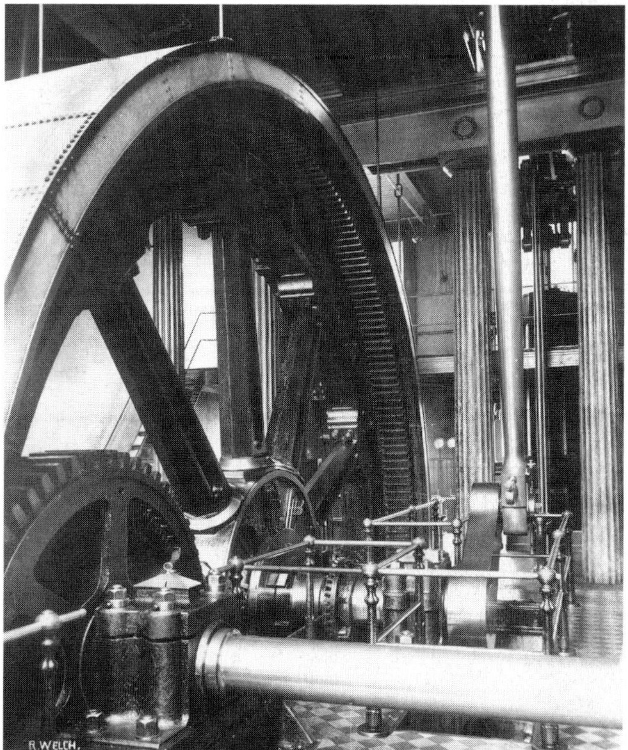

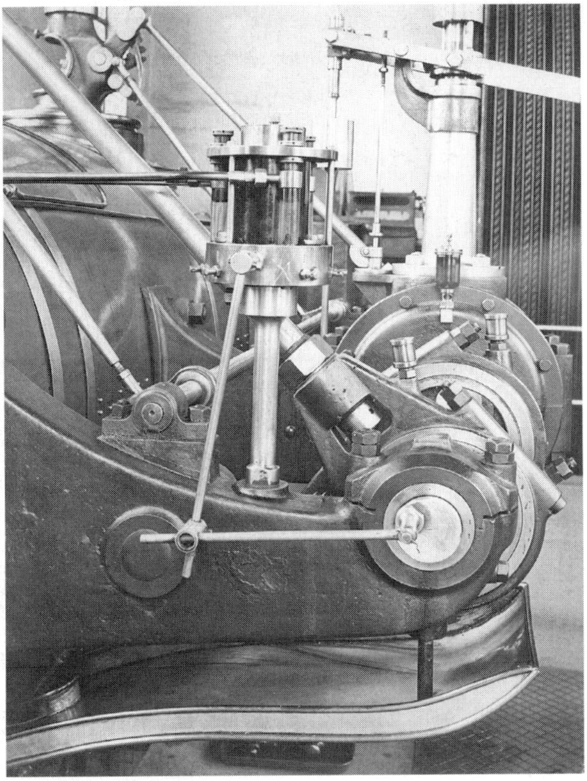

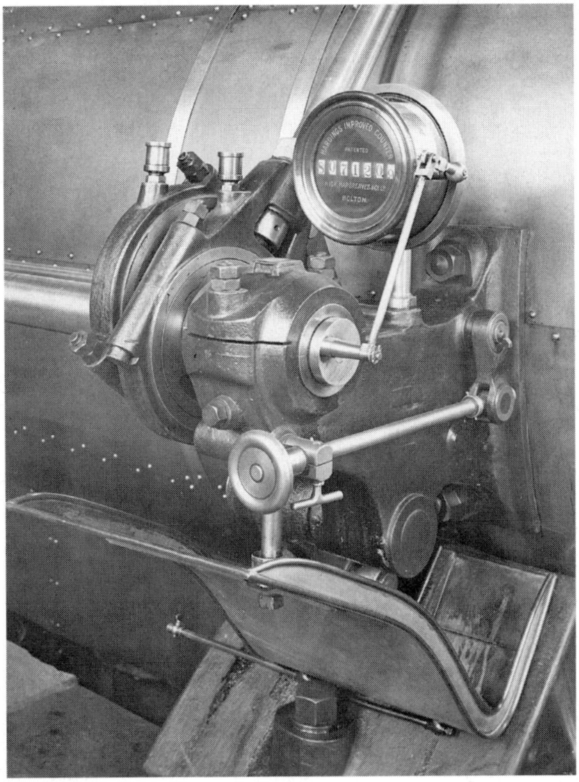
PATENTED
BOLTON

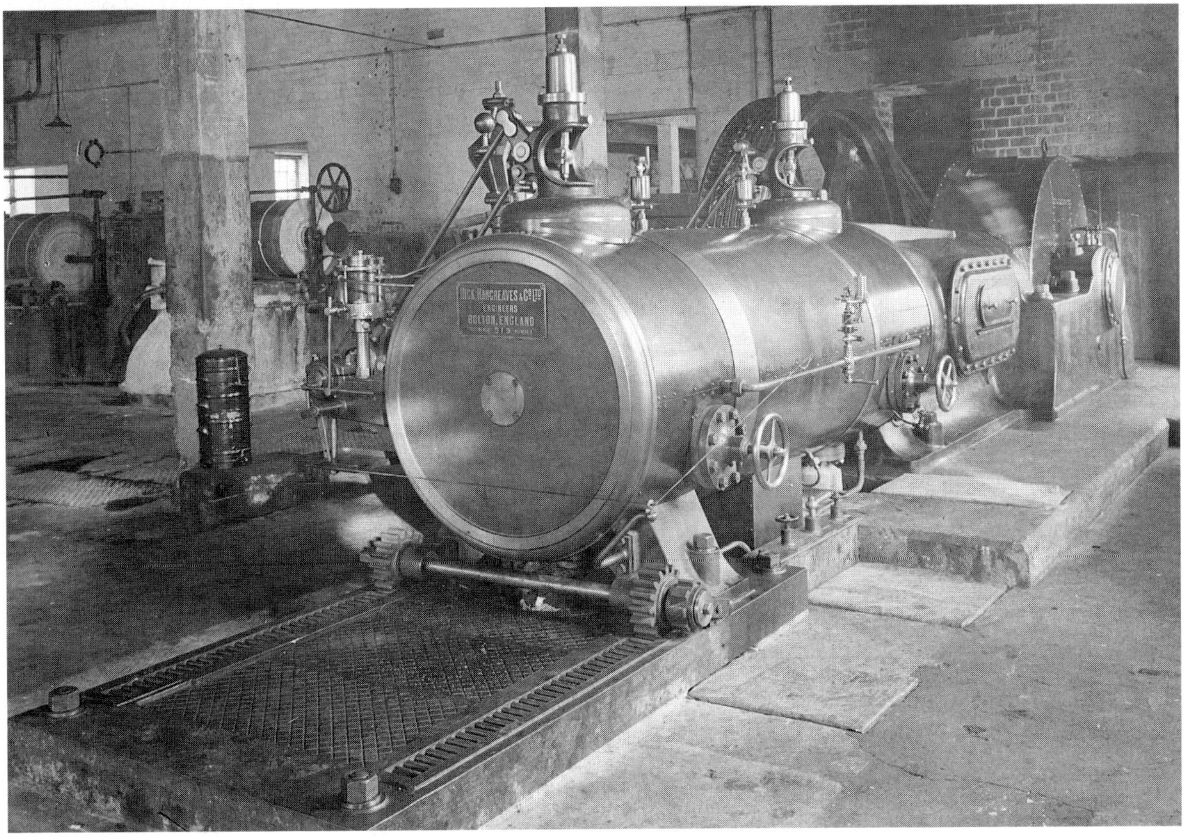
ENGINEERS
BOLTON, ENGLAND

MAK/165 - Higginbottom & Mannock Crown Ironworks, West Gorton, Manchester.

1) Advertisements:
1905, 1911.

MAK/166 - E. S. Hindley, Bourton, Dorset.

1) Advertisements:
1880, 1882, (2), 1887, 1891 (2), 1893, 1894 (2), 1895(2), 1896(2), 1896, 1897(2), 1901(2), 1905, 1908, 1910, 1913.

Reverse: 1882 Sandycroft Foundry & Engine Works, nr Chester.

Reverse: 1894 Wilson Hartnell, Volt Works, Leeds - electrical plant.

Reverse: 1894 Henderson & Glass, Liverpool - steel stockholder.

Reverse: 1895 London Emery Works, 12-10 Vine Street, Clerkenwell Road, London E. C.

Reverse: 1895 Clayton & Shuttleworth, Lincoln; Blackman Ventilating Co. Ltd, 63 Fore Street, London; Stanley Mathematical Instruments, Great Turnstile, Holborn, London W. C.

2) Letters:
Note by APW about Hindley engine in Sherborne museum.

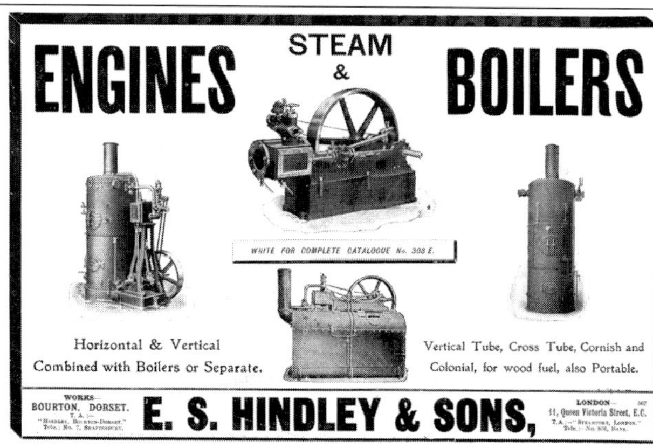

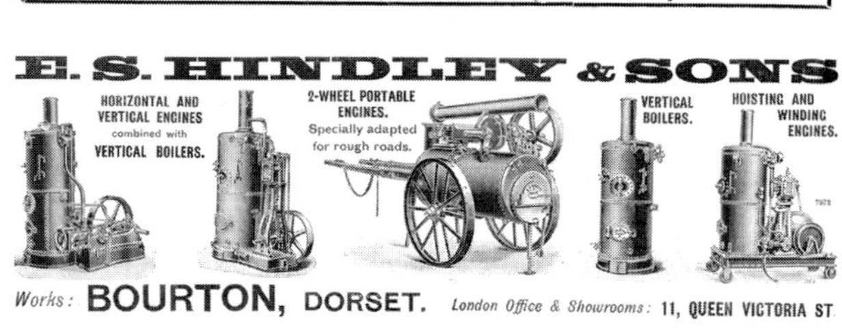

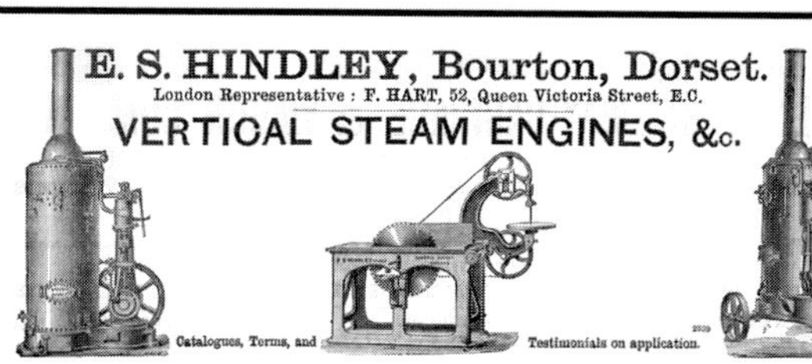

MAK/167 - Holborow & Co., Dudbridge Ironworks Stroud.

1) Advertisements:
Ads 1879, 1880, 1882, 1891, 1892, 1893.

Reverse: 1897 W. A. Martin, Pocock Street, Blackfriars Road, London S.E. - Smoke-consuming doors; Thomas Middleton & Co., Loman Street, Southwark - Brickmaking machinery.

5) Printed material:
Article from *Engineering,* 17 May 1878, describing a compound mill engine.

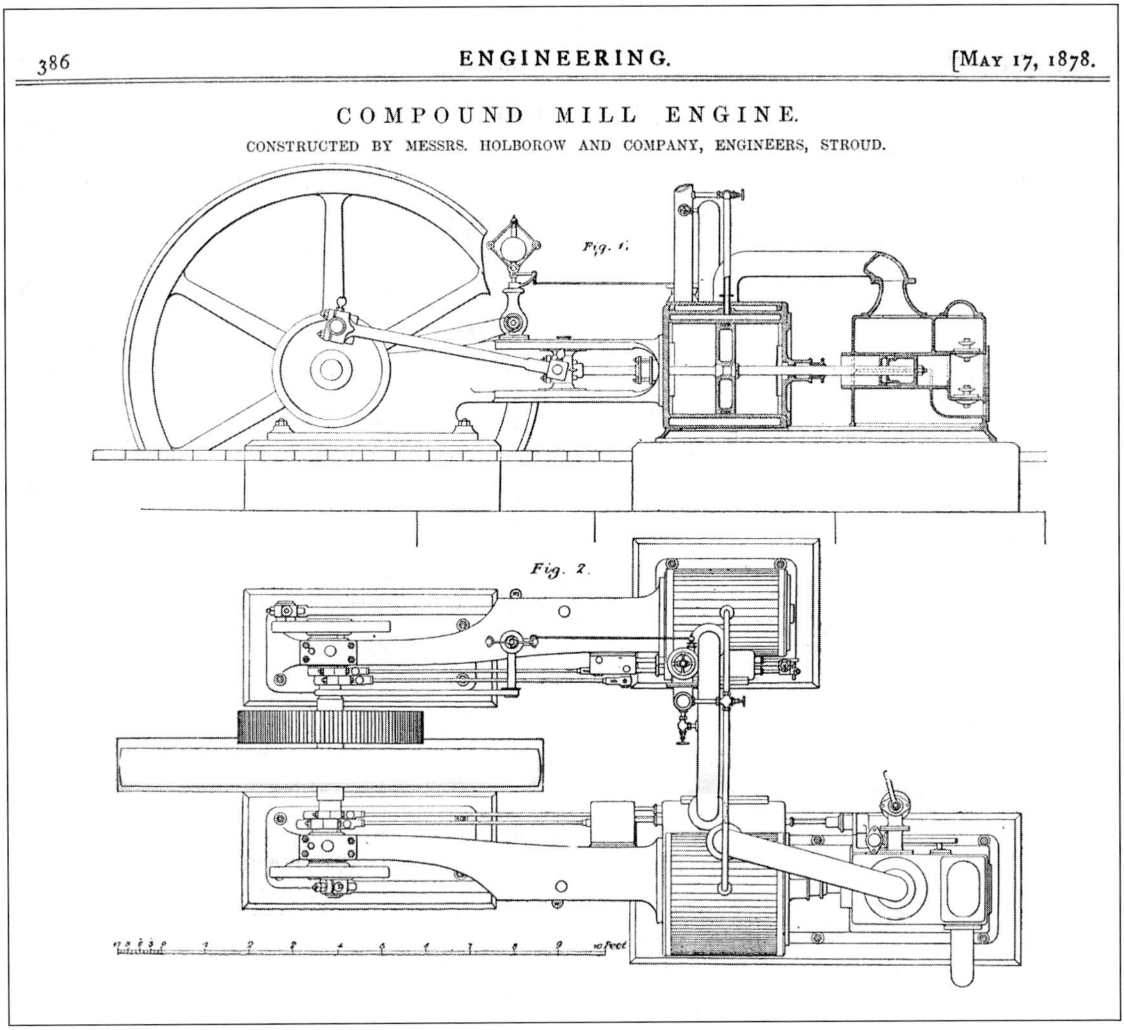
386 ENGINEERING. [MAY 17, 1878.

COMPOUND MILL ENGINE.

CONSTRUCTED BY MESSRS. HOLBOROW AND COMPANY, ENGINEERS, STROUD.

MAK/168 - Holdsworth & Sons, Croft Boiler Works, Bradford.

6) Trade literature:
Photocopy of booklet: *Yorkshire Boilers* n.d. but c 1920, 23pp.

MAK/169 - Hooven, Owens, Rentschler & Co., Hamilton, Ohio, U.S.A.

5) Printed material:
Article '*Sculptors in steel*', reprinted from *Scientific American*, Jan 1926 - Mainly about large machine tools but does have data about marine engines the firm built in WW1.

MAK/170 - Joseph Hopkinson, Huddersfield.

2) Letters:
Letters from Kirklees Museum to GW dated 12 and 29 August (no year noted) about supplying photographs and photocopies from Hopkinson's book.

3) MS material:
Set of MS notes by GW about Hopkinson's book (see below). 5pp.

4) Photographs:
Photograph of frontispiece to Hopkinson's book.

5) Printed material:
Photocopy of frontispiece and pp 278 - 285 from Joseph Hopkinson, *The working of the steam engine explained by the use of the indicator*, John Weale, 1863.

MAK/171 - W. H. Hooper & Co., Engineers, Moscow.

5) Printed material:
Plates from article from *The Engineer* 29 January 1897, describing a Corliss engine at the Nijni Novgorod exhibition - lacks the descriptive text.

MAK/172 - R. Hornsby & Sons, Spittlegate Iron Work, Grantham.

1) Advertisements:

Ads 1882, 1896 (2).

Reverse: 1882 Priestman Bros, Hull & London; Spence's Metal Manufacturing Co., 31 Lombard Street, London E.C.

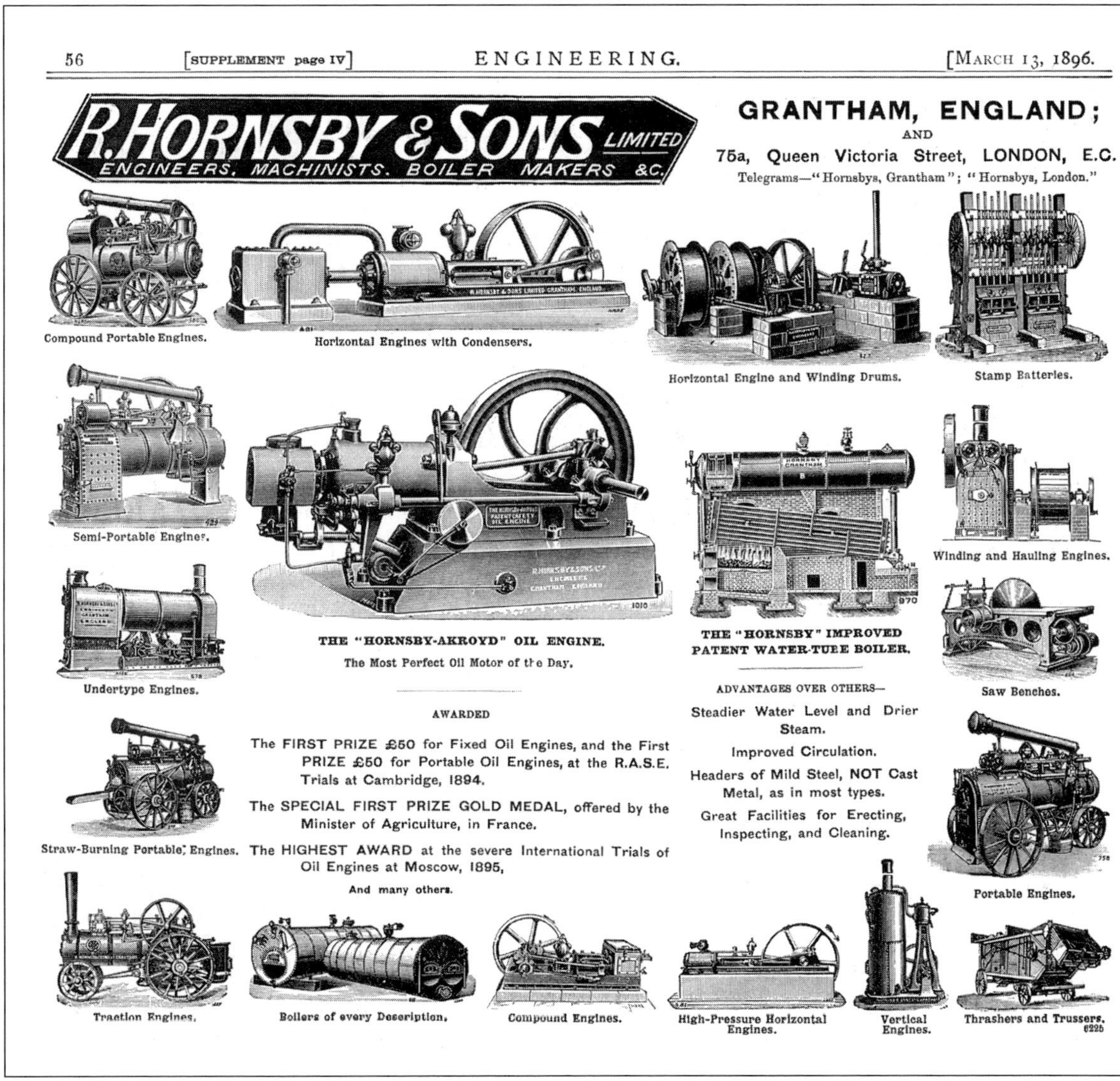

MAK/173 - Ralph Horsfield & Co., Chapel-en-le-Frith, Derbyshire.

1) Advertisements:
1894, 1897.

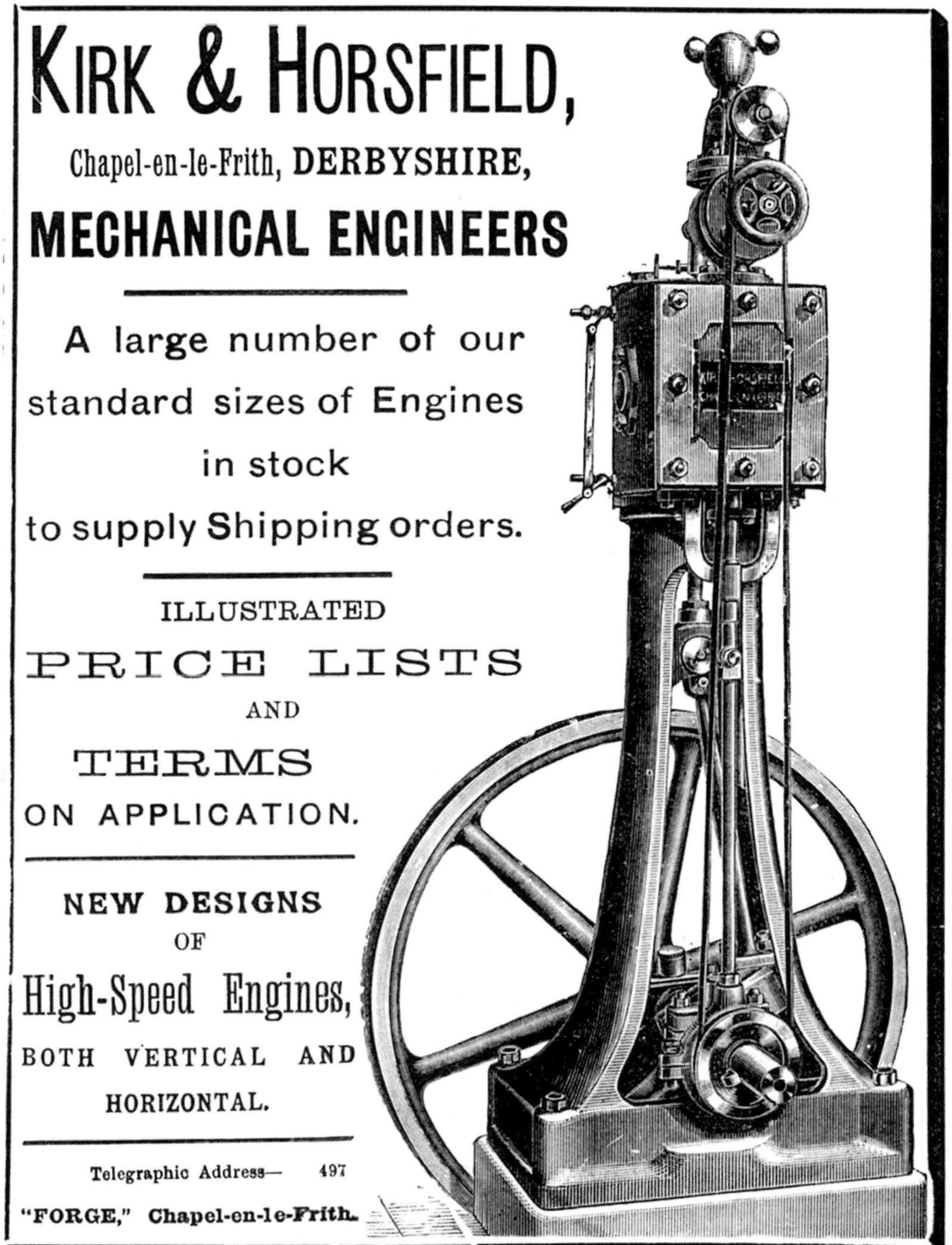

Ralph Horsfield & Co.

CHAPEL-EN-LE-FRITH,

DERBYSHIRE.

Telegrams - - "Forge, Chapel-en-le-Frith."

ESTABLISHED 1770.

"A B C" and "A 1" Codes used.

SPECIALISTS

IN THE MANUFACTURE OF

STEAM ENGINES

For EVERY DESCRIPTION of DRIVING and WINDING.

Vertical, Horizontal, Diagonal, Simple, Coupled, Compound, Condensing,

and every other type.

MANUFACTURED BY

Special Machinery,

Ensuring Accurate Workmanship. Efficient and Durable.

BOILERS

Lancashire, Cornish, Multitubular & Vertical.

1578

SEND FOR PRICE LISTS.

Estimates on Application.

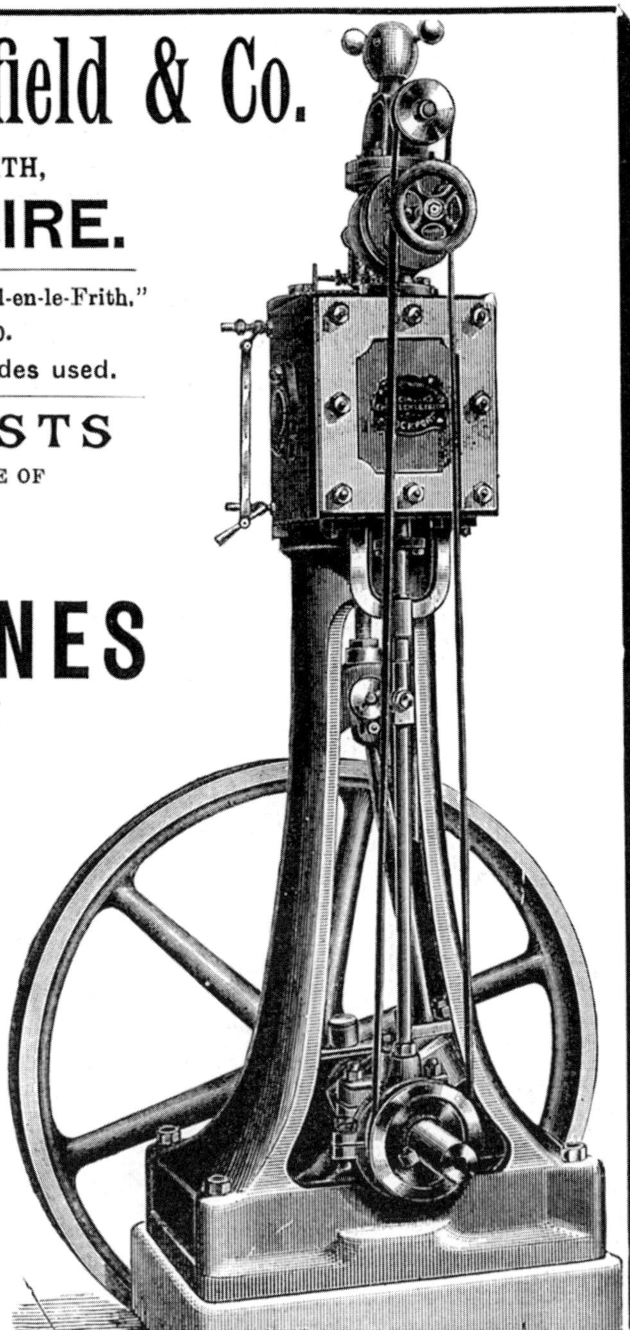

MAK/174 - J. & F. Howard, Bedford.

5) Printed material:

Copy of article from *Engineering* 3 March 1871 describing boilers of SS *Fairy Dell*, engines by Pollit & Wizgell.

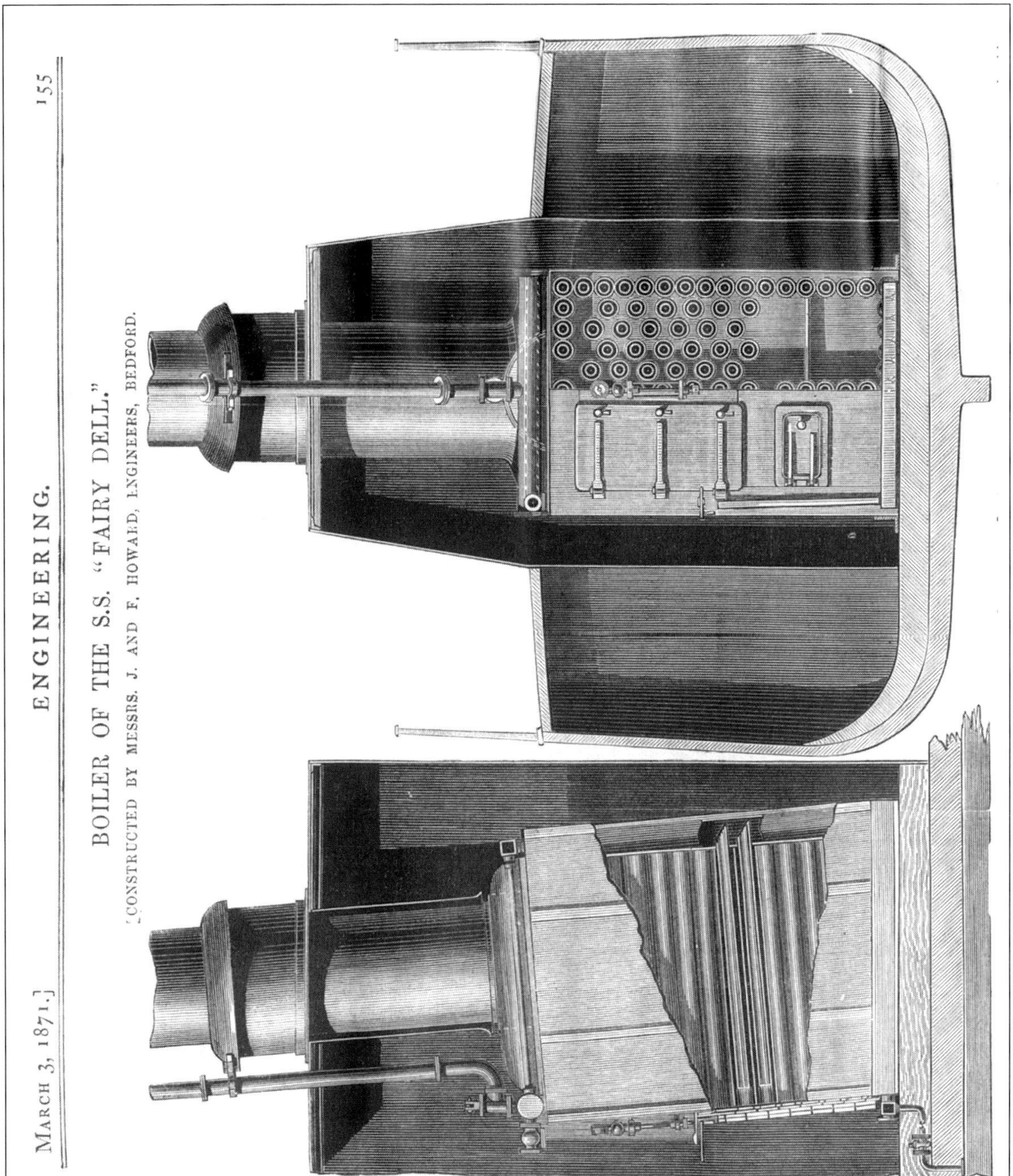

March 3, 1871.] ENGINEERING. 155

BOILER OF THE S.S. "FAIRY DELL."

[CONSTRUCTED BY MESSRS. J. AND F. HOWARD, ENGINEERS, BEDFORD.

MAK/175 - James Howden & Co., Scotland Street, Glasgow.

1) Advertisements:
Ads 1905 (2), 1908, 1910.

Reverse: 1905, Bruce Peebles & Co. Ltd, Edinburgh - PPP generators and motors.

Reverse: 1908 William Sellers, Philadelphia, Pennsylvania, USA.

Reverse: 1910 Werf Gusto, Firma A. F. Smulders, Schiedam, Holland - shipbuilders.

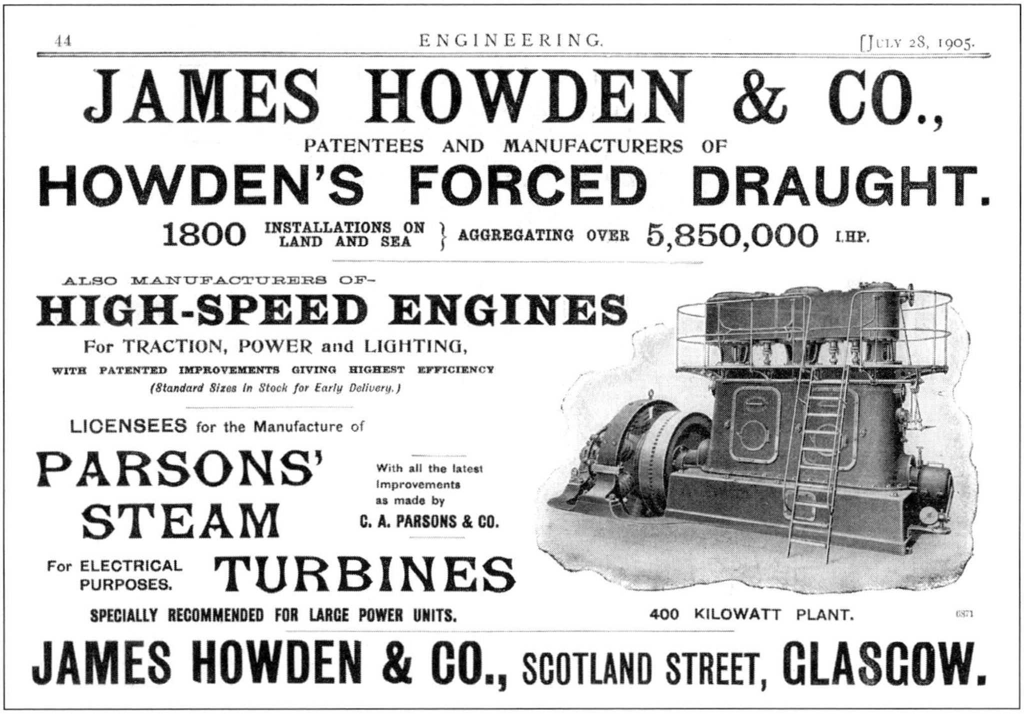

MAK/176 - E. Howell & Co., Ltd, Engineers, Poole, Dorset.

1) Advertisements:
1896 (2).

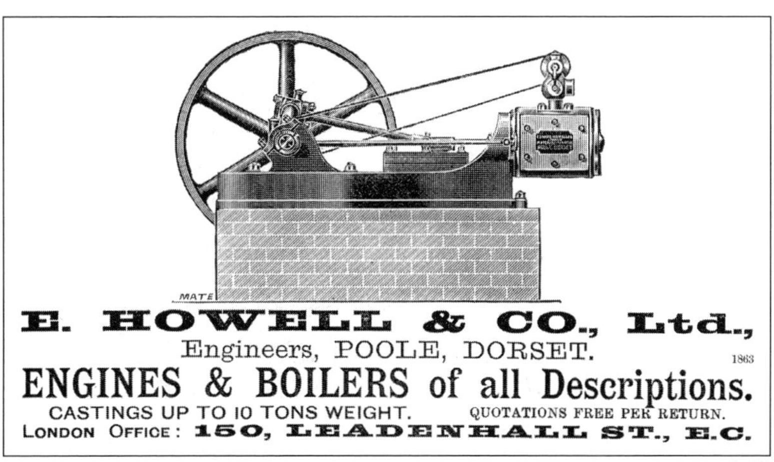

MAK/177 - Hulse & Co., Ordsal, Salford, Manchester.

1) Advertisements:
1887.

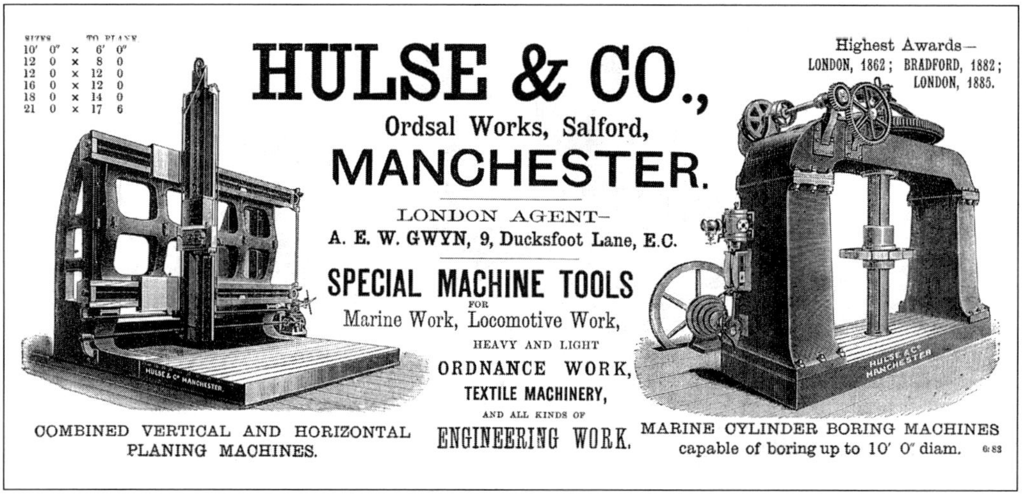

MAK/178 - Hunter & English, Bow, London.

5) Printed material:
Photocopy of article in *Engineering* 27 June 1879 about Hunter & Willans vertical compound.

MAK/179 - Hydraulic Engineering Co., Ltd, Chester.

4) Photographs:
Collection of 4 photographs of various engines, 1880, 1906, c 1910, 1911.

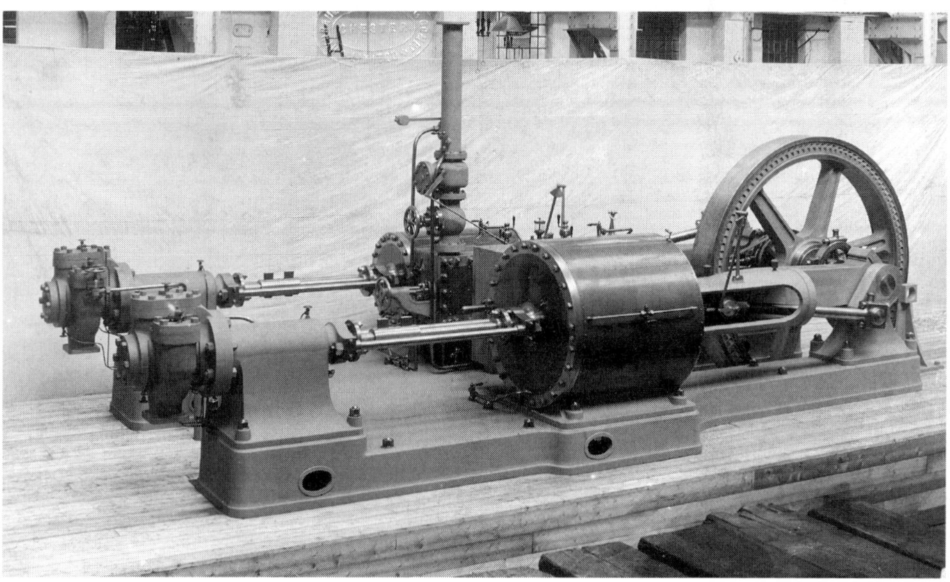

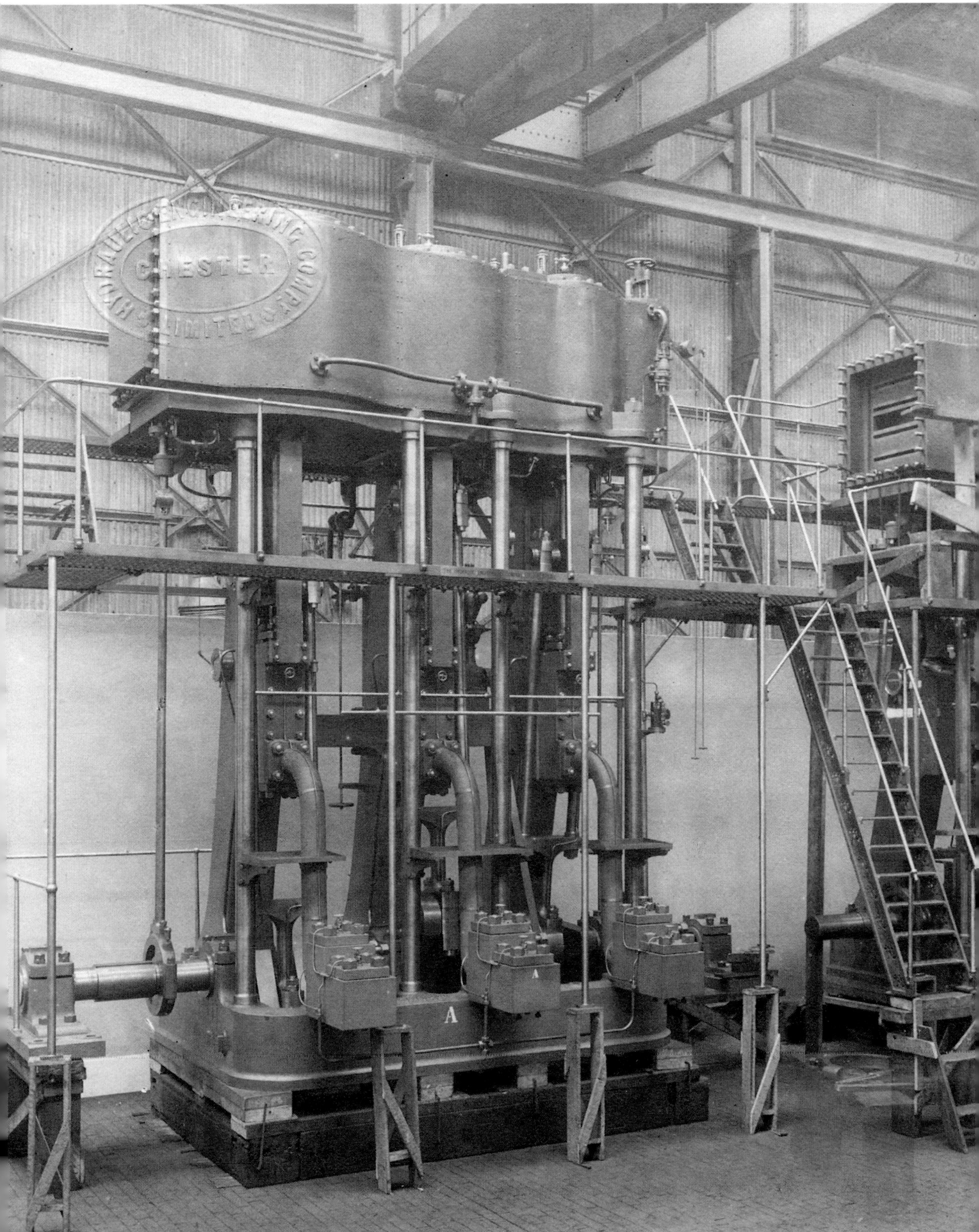
HYDRAULIC ENGINEERING COMPY
CHESTER
LIMITED
A

MAK/180 - I. H. C. Shipbuilders & Engineers, P. O. Box 6058, The Hague, Holland.

5) Printed material:
Photocopy of leaflet: *Semi-uniflow, 3 crank compound steam engine - I. H. C. Brouwer type*, c 1950, 3pp.

MAK/181 - T. James & Co., Albert Works, Great Suffolk Street, London, S.E.

1) Advertisements:
c 1870.

T. JAMES & Co.'s

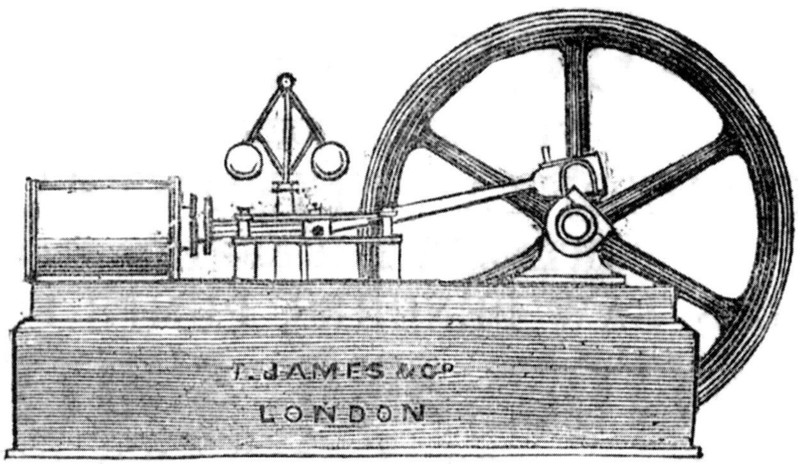

Improved Horizontal High Pressure Expansive Steam Engines.

3½-HP. diameter of cylinder, 6 in., stroke 10 in., price £25 10*s*.
Other sizes equally low. Price Lists on application.

Albert Works, Great Suffolk Street, London, S.E

MAK/182 - Jessop & Appleby Bros Ltd, London Steam Crane & Engine Works, Leicester.

5) Printed material:
Photocopy of pages 94-111 of Jessop & Appleby catalogue:, (see p. 172) 1899.

HORIZONTAL STEAM ENGINE

WITH OVERHUNG CYLINDER.

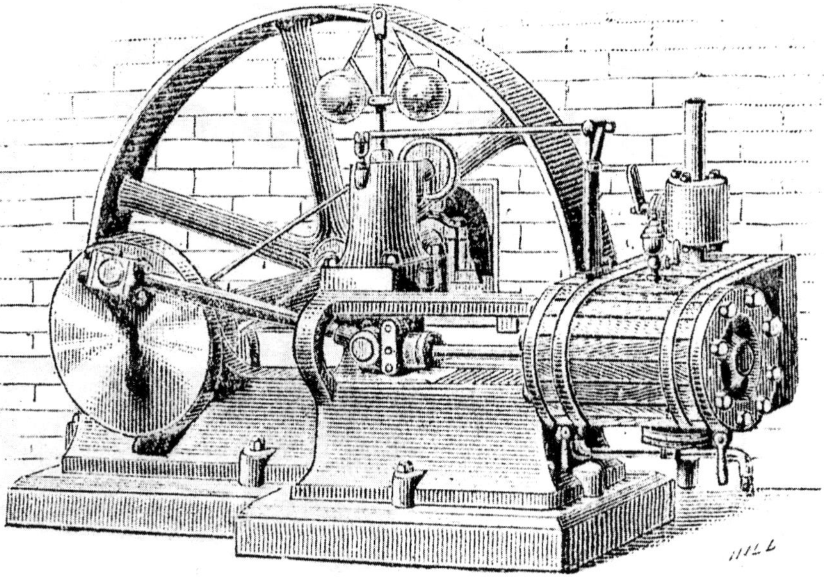

The crank shaft bearing, bored guides and steam cylinder are secured to a neat cast-iron bed plate as illustrated, and the Engine is complete with fly-wheel, governor, feed pumps, and pedestal, with wall box to carry the end of the crank-shaft ; the disc is balanced, turned bright, and fitted with a steel crank pin.

Prices of Horizontal Steam Engines.

NOMINAL HORSE POWER OF ENGINE	2	3	4	6
Diameter of cylinder .. ..inches	4	5	6½	8
Length of stroke .. .. ,,	7	9	10	14
Number of revolutions per minute .	200	200	150	125
Price of Engine, complete	£28	£38	£44	£68

JESSOP & APPLEBY BROS. LIMITED, Engineers, Leicester and London.

INDEX

For page references refer to the contents on page 5. This index covers both volumes.

Camerons, Manchester.	MAK/139
A. B. Cameron, Pump Works, 23rd St., New York, USA.	MAK/268
Campbell & Calderwood, Soho Engine Works, Paisley, Scotland.	MAK/279
Campbells & Hunter Ltd, Dolphin Foundry, Leeds.	MAK/232
Caramyl Manufacturing Co., 10 Marden Street, Manchester.	MAK/284
The Cariton Engineering Company, Works Glengall Road, Millwall, London.	MAK/357
J. Harrison Carter, Milling Engineer, 82 Mark Lane London, E. C.	MAK/232
Cedervall.	MAK/088
T. F. Chambers, Hull.	MAK/352
Alexander Chaplin & Co., Cranstonhill Engine Works, Govan, Glasgow.	MAK/160; /319
Chesters Engineering, Renfrew, Scotland.	MAK/070
Clarke, Chapman & Co., Gateshead on Tyne.	MAK/010; /326
Clarke's Crank Co., Patent Crank Works, Lincoln.	MAK/077
Henry Clayton, Son and Howlett, Atlas Works, Harrow Road, London.	MAK/369
Clayton, Howlett and Venables, Atlas Works, Harrow Road, London.	MAK/022
Clayton & Shuttleworth, Lincoln.	MAK/166
The Cleveland Bridge & Engineering Co., Darlington.	MAK/275
Clyde Rivet Works Co., Glasgow.	MAK/212; /258
Clyde Structural Iron Co., Ltd.	MAK/035
Cochran & Co., Annan.	MAK/064
David Colville & Sons	MAK/130; /156
The Consolidated Engineering Co., Ltd, Slough.	MAK/216
The Conveyor and Elevator Co., Accrington, Lancashire.	MAK/279
Co-operative Iron Works Company Ltd., North Moor Foundry, Oldham.	MAK/369
Cowans Sheldon, Carlisle.	MAK/026
Cox & Co., Falmouth Docks Iron Works, Falmouth, Cornwall.	MAK/128; /319
A. F. Craig, Caledonia Engine Works, Paisley, nr Glasgow	MAK/334
Craig & Donald, Johnstone, Nr Glasgow.	MAK/267
Crompton & Co.	MAK/360
Crosby Steam Gauge & Valve Co., 75 Queen Victoria Street, London.	MAK/319
Crossley Bros, Openshaw, Manchester .	MAK/056; /095
Crystal Palace Electrical Exhibition 1892.	MAK/285
D. M. Cumming, Blackhill Dock, Parkhead, Glasgow.	MAK/374
A. & W. Daglish, West of Scotland Boiler Works, Pollokshaws, Glasgow.	MAK/216
Darlington Forge Co., Ltd.	MAK/279
Davey Paxman, Colchester.	MAK/145; /255
Davis and Primrose, Etna Ironworks, Bangor Road, Leith N.B.	MAK/290
Davy Bros, Limited, Sheffield.	MAK/333
Dennystone Forge Co., Dumbarton.	MAK/235; /266
Walter Dixon & Co.	MAK/107
J. Donald & Son, 42 Cadogan Street, Glasgow.	MAK/339
Bryan Donkin, Bermondsey.	MAK/156
Douglas & Grant, Dunniker Foundry, Kirkcaldy.	MAK/369
Doulton & Co., Lambeth.	MAK/160
Downs & Thompson, Old Foundry, Hull.	MAK/208
Drysdale & Co., Bon Accord Engine Works, Glasgow.	MAK/037; /366
Richard Dudgeon, 114, Fenchurch Street, London E. C.	MAK/333
Ebbw Vale Steel & Iron and Coal Co.,	MAK/091
Thomas Edington & Sons, Phoenix Ironworks, Glasgow.	MAK/319
The Electric Constructional Company Ltd.	MAK/281
Fairbairn, Naylor, Macpherson & Co., Ltd., Wellington Foundry, Leeds.	MAK/366
The Falcon Engine & Car Works, Ltd., 41 Coleman Street, London and Loughborough.	MAK/009; /275

Hudswell, Clarke & Co., Ltd, Leeds.	MAK/053; /107; /203
Hughes & Lancaster, 47, Victoria Street, London S. W.	MAK/210
Hulme & Lund, Manchester.	MAK/290
R. Hunt & Co., Earls Colne, Essex.	MAK/255
Hunter & English, Engineers, Bow, London E.	MAK/360
George Hutchinson & Co., Forth Bank Oil Works, Newcastle upon Tyne.	MAK/223
Ibbotson Brothers & Co., Ltd., Globe Steel Works, Sheffield.	MAK/319
George Illiot & Co., 23, Great George Street, Westminster, London.	MAK/319
M. Immisch, Maiden Works, Prince of Wales Road, London NW.	MAK/075
Ingersoll-Rand.	MAK/088
C. Isler & Co., 88, Southwark Street, London S.E.	MAK/223
J. Jessop & Son, Leicester.	MAK/334
William Johnson & Sons, Milling & Brick Engineers, Castleton Foundry, Armley, Leeds.	MAK/073; /158; /266
Thomas Jowitt & Sons, Scotia Steel Works, Sheffield.	MAK/345
Kendall and Gent, Manchester.	MAK/056
Kennet Iron Works & Foundry, de Bohun Road, Reading.	MAK/317
Dick Kerr & Co., Ltd, London, Preston & Kilmarnock, N. B.	MAK/281
Kerr Stuart & Co., Light Railway Engineers & Contractors, 20 Bucklersbury, London.	MAK/319
Henry Kinsey, Robin Hood Works, Nottingham.	MAK/160
John Kirkcaldy, Engineer, 40 West India Dock Road, London.	MAK/290
The Klein Engineering Co., 94 Market Street, Manchester.	MAK/018
Conrad Knap & Co., London & Lincoln.	MAK/319
Körting Bros., 17 Lancaster Avenue, Manchester.	MAK/258
John H. Ladd & Co., 116 Queen Victoria Street, London.	MAK/290
Lancaster & Tonge, Pendleton.	MAK/142; /275
John Lang, Johnstone, nr Glasgow.	MAK/131
Laurence, Scott & Co., Gothic Works, Norwich.	MAK/001
J. W. Lee & Co., Queen Street Works, Leicester.	MAK/276
Leeming Webster, Carlisle Works, Dewsbury.	MAK/313
Leete, Edwards & Norman, Engineers, 366 & 368 Euston Road, London N. W.	MAK/311
Edwin Lewis & Son, Wolverhampton,	MAK/142; /160
Llewellins & James, Bristol,	MAK/275
Lobnitz & Co., Ltd. Engineers & Shipbuilders, Renfrew, Scotland,	MAK/312
London Emery Works, 12-10 Vine Street, Clerkenwell Road, London E. C.	MAK/056; /166
Lowood.	MAK/015
Luke & Spencer Ltd, Broadheath, Manchester.	MAK/103; /128
Lytham Shipbuilding and Engineering Co., Lytham.	MAK/011
P. & W. Maclellan, Ltd, 129 Trongate, Glasgow.	MAK/268
Magnolia Antifriction Metal Co., 49 Queen Victoria Street, London E. C.	MAK/255
Manning Wardle & Co., Boyne Engine Works, Leeds.	MAK/264
Edwin Mansfield & Co., 76 Queen Victoria Street, London E. C.	MAK/210
H. R. Marsden, Soho Foundry, Leeds.	MAK/300; /319
S. Marsden & Son, Ltd., London Road, Manchester.	MAK/312
John Marshall & Sons, Glasgow Tube Works.	MAK/256; / 279; /319
Marshall & Sons, Gainsborough.	MAK/067
W. A. Martin, Pocock Street, Blackfriars Road, London S. E.	MAK/167
Martin, Johnson and McKenzie, Engineers, 1, 3 & 5 Salthouse Lane, Salthouse Dock, Liverpool.	MAK/369
B. S. Massey,Openshaw, Manchester.	MAK/067

Thomas McCulloch & Sons, Kilmarnock.	MAK/319
McKie & Baxter, Copland Works, Govan, Glasgow.	MAK/279
Charles McNeil, jr., Kinning Park Ironworks, 270, West Scotland Street, Glasgow.	MAK/361
McPhail & Simpson, Engineers, Wakefield.	MAK/217
Measure Bros Ltd., 53b Southwark Street, London and Pitlake, Croydon.	MAK/109; /312; /333
Mellowes & Co., Ltd, Sheffield.	MAK/232
W. M. Melville, Sinclairtown Foundry, Kirkcaldy.	MAK/127
The Mersey Steel and Iron Company, Liverpool.	MAK/352
Metallic Valve Co., Tower Buildings, Water Street, Liverpool.	MAK/015
La Meuse, Liège, Belgium.	MAK/074
Robert Middleton, Foundry, Leeds.	MAK/255
Thomas Middleton & Co., Loman Street, Southwark.	MAK/167
Mitchell's Emery Wheel Co., Mill Street, Bradford, Manchester.	MAK/300
Richard Morland & Sons, 3 Old Street, London E. C.	MAK/156; /319
Alexander Morton & Thomson, 96 Buchanan Street, Glasgow.	MAK/272
Francis Morton & Co., Naylor Street, Liverpool.	MAK/344
William Muir & Co., Sherborne Street, Manchester.	MAK/361
A. G. Mumford Ltd, Colchester.	MAK/232
Wm. Murdoch 30 Hope Street, Glasgow.	MAK/253
John Musgrave & Sons, Bolton.	MAK/102
Nasmyth, Wilson & Co., Ltd, Bridgewater Foundry, Patricroft, Manchester.	MAK/347
De Negri and Herrman, Engineers, Hornsey Iron Works, Blenheim Road, Hornsey Road, Upper Holloway, London.	MAK/369
Newton & Nicholson, Tyne Dock Corrugated Packing Works, South Shields.	MAK/279
John Okey & Sons, 172, Blackfriars Road, London S.	MAK/277
Alexander Oldham & Sons, Dunkinfield, nr Manchester.	MAK/327
The Oldham Boiler Works Co., Ltd.	MAK/056
S. Oppenheim & Co.,	MAK/191
S. Owens & Co., Whitefriars Street, London EC,	MAK/108
C. A. Parsons, Heaton Works, Newcastle upon Tyne.	MAK/052; /156
Partington's, Bradford, Yorks.	MAK/255
Patent Rivet Co., Ltd., Smethwick.	MAK/248
Patent Steam Boiler Co., Heneage Street, Birmingham.	MAK/077
Frank Pearn & Co., Ltd, West Gorton, Manchester.	MAK/275
D. Bruce Peebles & Co., Tay Works, Edinburgh.	MAK/185; /360
Penman & Co., Glasgow.	MAK/102
J. W. & C. Phillips, 23 College Hill, London, E. C.	MAK/210
Phoenix Bolt & Nut Co., Handsworth, Birmingham.	MAK/088
Thomas Piggot & Co., Atlas Works, Birmingham.	MAK/062; /107; /139; /241
Samuel Platt, Wednesbury.	MAK/319
Pott, Cassels & Williamson, Motherwell, nr Glasgow.	MAK /300
Price's Patent Candle Co., Ltd, Belmont Works, Battersea, London S. W.	MAK/374
Priestman Bros, Hull & London.	MAK/009; /173
Pulsometer Eng. Co., Nine Elms Iron Works, London, S. W. and Reading.	MAK/067; /357
Ransome, Sims & Jefferies, Ipswich.	MAK/070
F. Reddaway, Pendleton, Manchester.	MAK/139
Geo. Richards & Co., Ltd.	MAK/128
Ropeways Syndicate, 150 Leadenhall Street, London.	MAK/142
Rose, Downs and Thompson, Old Foundry, Hull.	MAK/001; /235; /266

B. R. Rowland & Co., Ltd, Climax Works, Reddish, nr Manchester. MAK/279
Rugby Portland Cement Co., Rugby, Warwickshire. MAK/352
Ruston Proctor & Co., MAK/035

Sandycroft Foundry & Engine Works, nr Chester. MAK/166; /262
Saville Street Foundry and Engineering Works, Sheffield. MAK/054; /131; /157
George Scott & Son, London. MAK/109
Scott Bros. West Mount Ironworks, Halifax. MAK/319
Scriven & Co., Leeds. MAK/156
Selig, Sonnenthal & Co., 85 Queen Victoria Street and Lambeth Hill, London. MAK/070; /132
William Sellers, Philadelphia, Pennsylvania, USA. MAK/175
Thomas Shanks, Union Iron Works, Glasgow. MAK/010
Sherwin Feedwater Heaters, Almea Street, Birmingham. MAK/210
T. Shore & Sons, Albion Foundry, Etruria, Stoke on Trent. MAK/266
W. Simons & Co., Shipbuilders, Renfrew, nr Glasgow. MAK/319
James Simpson, Newark on Trent. MAK/063
Slack's Emery Wheel and Machine Co., Ltd, Lime Bank Works, Ashton Old Road, Manchester. MAK/157; /333
Smith & Grace, Thrapston MAK/107
John Smith & Co., Grove Works, Carshalton, Surrey. MAK/255
Spencer & Gillett, Melksham, Wilts. MAK/275
Werf Gusto, Firma A. F. Smulders, Schiedam, Holland. MAK/175
Spence's Metal Manufacturing Co., 31 Lombard Street, London E. C. MAK/173
John Spencer & Sons Ltd, Newburn Steel Works, Newcastle upon Tyne. MAK/214
Standard Ammonia Co., Ltd., Old Ford, London E. MAK/25
Stanley Mathematical Instruments, Great Turnstile, Holborn, London W. C.MAK/166
The Steel Company of Scotland, 150, Hope Street, Glasgow, and 23 Royal Exchange Square, Glasgow. MAK /326; /352; /354
The Steel Pipe Co., Kirkaldy, Scotland. MAK/222; /279
L. Sterne & Co., Crown Ironworks, Glasgow. MAK/011; /280
Stevens & Struthers, Glasgow. MAK/214
D. Stewart & Co., (1902) Ltd, London Road Iron Works, Glasgow. MAK/011; /275
D. Y. Stewart & Co., Glasgow. MAK/374
The Stirling Boiler Co., Ltd, 25 Victoria Street, Westminster, S. W. MAK/314
Stothert & Pitt, Bath. MAK/127
T. Sugden Ltd., 180 Fleet Street, London. MAK/363

The New Taite Howard Pneumatic Tool Co., Ltd. 63 Queen Victoria Street, London E. C. MAK/350
Tangyes Ltd, Birmingham. MAK/232
Geo. Thomas & Co., 28 Deansgate, Manchester. MAK/284
Thomson & Sterne, Crown Iron Works, Glasgow. MAK/022
John I. Thornycroft & Co., Chiswick, London W. MAK/319
Thwaites Brothers, Vulcan Ironworks, Bradford. MAK/157
H. Titley & Son, Alton Works, Soho, Smethwick. MAK/077
J. B. Treasure & Co., 8 Vauxhall Road, Liverpool. MAK/290
George Turton, Patent Buffer, Steel and File works, Sheffield. MAK/319
Thomas Turton & Sons, Ltd, Sheaf Works and Spring Works, Sheffield. MAK/290; /319
Tube Company, 100c Queen Victoria Street, London. E. C. MAK/361

The Union Engineering Co., Ltd, Sophia Street, Rochdale Road, Manchester. MAK/319
The United Asbestos Co., Ltd., 161 Queen Victoria Street, London, E. C. MAK/319
United Kingdom Lighting Trust Ltd, 231 Strand London W. C. MAK/244
United States Metallic Packing Co., Ltd, Bradford. MAK/265
Urquhart, Lindsay & Co., Blackhouse Foundry, Dundee. MAK/011

The Vacuum Oil Company, The Albany, Victoria Street, Westminster, London S.W.	MAK/319
Vaughan & Sons, Ltd, Manchester.	MAK/032
A. Verey, Buckland Engine Works, Dover.	MAK/145
Walker Campbell & Co., 59, Bridgewater Street, Liverpool.	MAK/361
Henry Wallwork & Co., Charter Street, Manchester.	MAK/285
T. & E. Wannitracher, 28a Basinghall Street, London E. C.	MAK/312
H. W. Ward, 89 Lionel Street, Birmingham.	MAK/131
Robert Warner & Co., 97 Queen Victoria Street, London. and Walton on the Naze, Essex.	MAK/010; /290
Watson, Laidlaw & Co.,.	MAK/232
Weardale Iron & Coal Co., Tudhoe Ironworks, Spennymoor, Durham.	MAK/156; /258
Webster & Bennett, Machine Tool Makers, Coventry.	MAK/107
G. & J. Weir, 49 Jamaica Street, Glasgow.	MAK/339
The Weldless Steel Tube Co., Icknield Port Road, Birmingham.	MAK/328
Westinghouse Electric Co., Ltd, 4 Victoria Mansions, 32 Victoria Street, London S. W. 6.	MAK/220
W. Whittam, Salford, Manchester.	MAK/288
John H. Widdowson, Britannia Works, Ordsall Lane, Salford, Manchester.	MAK/308
W. H. Wilcox & Co., 34 & 36 Southwark Street, London S. E.	MAK/262
J. Wilkinson & Sons, Saltley, Birmingham.	MAK/266
Willans and Robinson Ltd., Engineers, Thames Ditton, Surrey.	MAK/356
Alex Wilson, Vauxhall Ironworks,	MAK/139
Wilson Brothers & Co., Engineers & lronfounders, Darlington.	MAK/334
Wilsons and Union Tube Co., Ltd, Govan and Union Tube Works, Coatbridge.	MAK/266
The Wire Tramway Co., 72, Mark Lane, London W. C.	MAK/262
A. Wood & Sons, 8 & 10 Stockwell Street, Glasgow.	MAK/279
Woodhouse & Mitchell, Brighouse, Yorks	MAK/366
"Woodite" and "Whaleite" Works, Mitcham Common, Surrey.	MAK/210
Woods, Cocksedge & Warner, Engineers, Stowmarket.	MAK/320
Yarrow, London.	MAK/010; /211